SIMULATION TECHNIQUES IN FINANCIAL RISK MANAGEMENT

Second Edition

WILEY SERIES IN STATISTICS IN PRACTICE

Advisory Editor, MARIAN SCOTT, *University of Glasgow, Scotland, UK*

Founding Editor, VIC BARNETT, *Nottingham Trent University, UK*

Statistics in Practice is an important international series of texts which provide detailed coverage of statistical concepts, methods, and worked case studies in specific fields of investigation and study.

With sound motivation and many worked practical examples, the books show in down-to-earth terms how to select and use an appropriate range of statistical techniques in a particular practical field within each title's special topic area.

The books provide statistical support for professionals and research workers across a range of employment fields and research environments. Subject areas covered include medicine and pharmaceutics; industry, finance, and commerce; public services; the earth and environmental sciences, and so on.

The books also provide support to students studying statistical courses applied to the above areas. The demand for graduates to be equipped for the work environment has led to such courses becoming increasingly prevalent at universities and colleges.

It is our aim to present judiciously chosen and well-written workbooks to meet everyday practical needs. Feedback of views from readers will be most valuable to monitor the success of this aim.

A complete list of titles in this series appears at the end of the volume.

SIMULATION TECHNIQUES IN FINANCIAL RISK MANAGEMENT

Second Edition

NGAI HANG CHAN AND HOI YING WONG
The Chinese University of Hong Kong

Library of Congress Cataloging-in-Publication Data:

Chan, Ngai Hang.
  Simulation techniques in financial risk management / Ngai Hang Chan and Hoi Ying Wong. – Second
edition.
      pages cm. – (Statistics in practice)
   Includes bibliographical references and index.
  ISBN 978-1-118-73581-7 (hardback)
  1. Finance–Simulation methods. 2. Risk management–Simulation methods. I. Wong, Hoi Ying,
1974- II. Title.
  HG173.C47 2015
  338.5–dc23
                                                                                                2015001921

Cover image courtesy of iStockphoto © pawel.gaul

Typeset in 10/12pt TimesLTStd by Laserwords Private Limited, Chennai, India

Printed in the United States of America

10 9 8 7 6 5 4 3 2 1

2   2015

To our families
N.H. Chan and H.Y. Wong

CONTENTS

LIST OF FIGURES

LIST OF TABLES

PREFACE

PREFACE TO THE SECOND EDITION

This book has now been in print for almost 10 years and has seen several printings. During this period, the field of quantitative finance has experienced abrupt changes, some for better and some for worse. But it has been very gratifying to us to have heard from many readers that this book has been helpful to them in dealing with the ever-changing financial landscape. It appears that to some extent at least the original objectives set out in the first edition have been realized. This book can be used either as an introductory text to simulations at the senior undergraduate or as a Master's level course. It can also be used as a complimentary source to the more specialized treatise by Chan and Wong (2013) entitled *Handbook of Financial Risk Management: Simulations and Case Studies*.

This second edition has been thoroughly revised and enhanced. Many of these changes were results of teaching different courses in simulation for financial risk managers over the years. In addition to cleaning up as many errors and misprints as possible, the following specific changes have been incorporated in this revision.

- Many readers suggested more exercises with worked solutions. As a result, we enlarge the problems and answers section in light of these requests.
- Because the use of VBA in Excel has been common in the financial industry, the current edition incorporates this suggestion. We have now replaced all S-Plus codes with VBA codes.
- Due to the advent in IT technology, a new website has been set up for readers to download the VBA computer codes.
  http://www.sta.cuhk.edu.hk/Book/SRMS/

As long as the website is available, we no longer print computer codes, so that more space can be used for expanded topics.

- Likewise, suggested solutions to exercises at the end of each chapter are now available via online supplementary materials.
- To make the book self-contained, two new chapters, Chapters 1 and 2, have been added. Chapter 1 introduces basic concepts of Excel VBA, and Chapter 2 introduces basic concepts of derivatives.
- Corresponding to Chapter 9 in the first edition, Chapter 11 of this edition is expanded to discuss in detail a one-factor interest rate model and the calibration to yield curves.
- More examples have been added to illustrate the concept of MCMC, in particular the Metropolis–Hastings algorithm.

Finally, we would like to thank colleagues and students alike, who have been giving us suggestions and ideas throughout the years. In particular, we would like to thank the editorial assistance of Dr. Warwick Yuen and Mr. Tom Ng of CUHK and Ms. Sari Friedman and Mr. Jon Gurstelle of Wiley. We also want to express our gratitude to the Research Grants Council of HKSAR for support at various stages of our work on this revision.

NGAI HANG CHAN AND HOI YING WONG

Shatin, Hong Kong

PREFACE TO THE FIRST EDITION

Risk management is an important subject in finance. Despite its popularity, risk management has a broad and diverse definition that varies from individual to individual. One fact remains, however. Every modern risk management method comprises a significant amount of computations. To assess the success of a risk management procedure, one has to rely heavily on simulation methods. A typical example is the pricing and hedging of exotic options in the derivative market. These over-the-counter options experience very thin trading volume, and yet their nonlinear features forbid the use of analytical techniques. As a result, one has to rely on simulations in order to examine their properties. It is therefore not surprising that simulation has become an indispensable tool in the financial and risk management industry today.

Although simulation as a subject has a long history by itself, the same cannot be said about risk management. To fully appreciate the power and usefulness of risk management, one has to acquire a considerable amount of background knowledge across several disciplines: finance, statistics, mathematics, and computer science. It is the synergy of various concepts across these different fields that marks the success of modern risk management. Although many excellent books have been written on the subject of simulation, none has been written from a risk management perspective. It is therefore timely and important to have a text that readily introduces the modern techniques of simulation and risk management to the financial world.

This text aims at introducing simulation techniques for practitioners in the financial and risk management industry at an intermediate level. The only prerequisite is a standard undergraduate course in probability at the level of Hogg and Tanis (2006), say, and some rudimentary exposure to finance. The present volume stems from a set of lecture notes used at the Chinese University of Hong Kong. It aims at striking a balance between theory and applications of risk management and simulations, particularly along the financial sector. The book comprises three parts.

- Part one consists of the first three chapters. After introducing the motivations of simulation in Chapter 1, basic ideas of Wiener processes and Itô's calculus are introduced in Chapters 2 and 3. The reason for this inclusion is that many students have experienced difficulties in this area because they lack the understanding of the theoretical underpinnings of these topics. We try to introduce these topics at an operational level so that readers can immediately appreciate the complexity and importance of stochastic calculus and its relationship with simulations. This will pave the way for a smooth transition to option pricing and Greeks in later chapters. For readers familiar with these topics, this part can be used as a review.
- Chapters 4–6 comprise the second part of the book. This part constitutes the main core of an introductory course in risk management. It covers standard topics in a traditional course in simulation, but at a much higher and succinct level. Technical details are left in the references, but important ideas are explained in a conceptual manner. Examples are also given throughout to illustrate the use of these techniques in risk management. By introducing simulations this way, both students with strong theoretical background and students with strong practical motivations get excited about the subject early on.

- The remaining Chapters 7–10 constitute part 3 of the book. In this part, more advanced and exotic topics of simulations in financial engineering and risk management are introduced. One distinctive feature in these chapters is the inclusion of case studies. Many of these cases have strong practical bearings such as pricing of exotic options, simulations of Greeks in hedging, and the use of Bayesian ideas to assess the impact of jumps. By means of these examples, it is hoped that readers can acquire a first-hand knowledge about the importance of simulations and apply them to their work.

Throughout the book, examples from finance and risk management have been incorporated as much as possible. This is done throughout the text, starting at the early chapter that discusses VaR of Dow to pricing of basket options in a multiasset setting. Almost all of the examples and cases are illustrated with Splus and some with Visual Basics. Readers would be able to reproduce the analysis and learn about either Splus or Visual Basics by replicating some of the empirical work.

Many recent developments in both simulations and risk management, such as Gibbs sampling, the use of heavy-tailed distributions in VaR calculation, and principal components in multiasset settings are discussed and illustrated in detail. Although many of these developments have found applications in the academic literature, they are less understood among practitioners. Inclusion of these topics narrows the gap between academic developments and practical applications.

In summary, this text fills a vacuum in the market of simulations and risk management. By giving both conceptual and practical illustrations, this text not only provides an efficient vehicle for practitioners to apply simulation techniques, but also demonstrates a synergy of these techniques. The examples and discussions in later chapters make recent developments in simulations and risk management more accessible to a larger audience.

Several versions of these lecture notes have been used in a simulation course given at the Chinese University of Hong Kong. We are grateful for many suggestions, comments, and questions from both students and colleagues. In particular, the first author is indebted to Professor John Lehoczky at Carnegie Mellon University, from whom he learned the essence of simulations in computational finance. Part 2 of this book reflects many of the ideas of John and is a reminiscence of his lecture notes at Carnegie Mellon. We would also like to thank Yu-Fung Lam and Ka-Yung Lau for their help in carrying out some of the computational tasks in the examples and for producing the figures in LaTeX, and to Mr. Steve Quigley and Ms. Susanne Steitz, both from Wiley, for their patience and professional assistance in guiding the preparation and production of this book. Financial support from the Research Grant Council of Hong Kong throughout this project is gratefully acknowledged. Last, but not least, we would like to thank our families for their understanding and encouragement in writing this book. Any remaining errors are, of course, our sole responsibility.

NGAI HANG CHAN AND HOI YING WONG

Shatin, Hong Kong

1

PRELIMINARIES OF VBA

1.1 INTRODUCTION

This chapter introduces the elementary programming skills in Visual Basic for Applications (VBA) that we use for numerical computation of the examples in the book. Experienced readers can read this chapter as a quick review.

1.2 BASIS EXCEL VBA

Microsoft Excel is widely used in the financial industry for performing financial calculations. VBA is a common programming language linked to Excel and other Microsoft Office software that was developed to automatically control and perform repetitive actions. In this section, we guide readers on how to start a VBA in Microsoft Excel and give some popular algorithms for performing repetitions. In most cases, simple algorithms will be sufficient to perform the computations in the examples and exercises. We provide the illustrations in Excel 2010, although other versions can be set up in a similar way. For a comprehensive reference, readers are referred to other books.

Simulation Techniques in Financial Risk Management, Second Edition. Ngai Hang Chan and Hoi Ying Wong.
© 2015 John Wiley & Sons, Inc. Published 2015 by John Wiley & Sons, Inc.

1.2.1   Developer Mode and Security Level

For first-time users of VBA in Excel, it is more convenient to switch on the developer mode, where many of the VBA functions can be easily accessed. To open the developer mode, follow the following steps:

Click [File] → [Options] (Fig. 1.1) → [Customize Ribbon] (Fig. 1.2) →
[Developer].

Figure 1.3 shows the ribbons at the top of Excel after switching on the developer mode. Macros refer to the codes executed in the VBA language. To execute the macros promptly, users are recommended to turn down the security level as follows:

Click [Macro Security] (Fig. 1.3) → Macro Settings [Enable all macros] (Fig. 1.4).

1.2.2   Visual Basic Editor

To edit the VBA codes, Microsoft provides a Visual Basic editor (VBE) in Excel for editing the macros. Macros are created, edited, and debugged in the VBE before being executed. A macro is usually created as a *Sub* or *Function* procedure that can perform automatic tasks, while a module consists of one or more macros. Similarly, a project has one or more modules. *Sub* and *Function* are reserved keywords in VBA. Users need to avoid using keywords when defining new variables. The codes in the

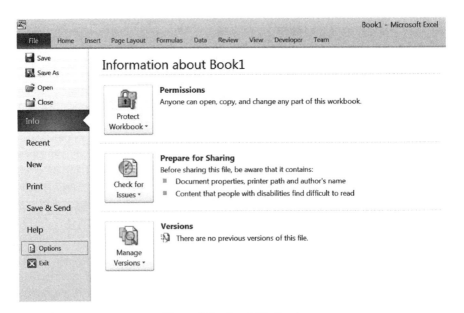

Figure 1.1   Excel [Options].

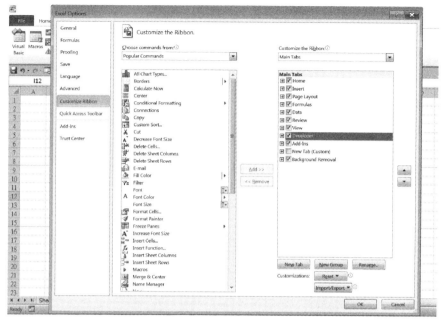

Figure 1.2   Developer mode selection.

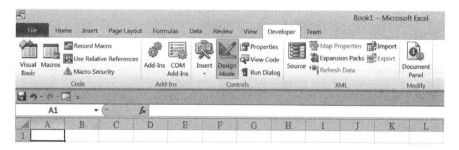

Figure 1.3   Excel in developer mode.

VBE are saved together with the Excel worksheet. In Excel 2010, these worksheets can be saved as .xlsm as an Excel Macro-Enabled Workbook file. To open and edit macros in VBE, follow the following procedure:

1. Open VBE: click the [Visual Basic] button under the developer mode (Fig. 1.3) or press ALT+F11.
2. Insert module: in the project window of the VBE, right-click on one of the worksheets, and select [Insert] → [Module] (Fig. 1.5).
3. Edit in VBE: type the codes in the panel on the right.

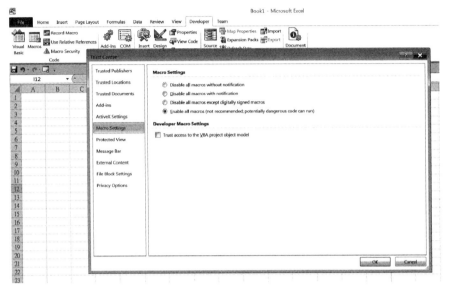

Figure 1.4   Macro security.

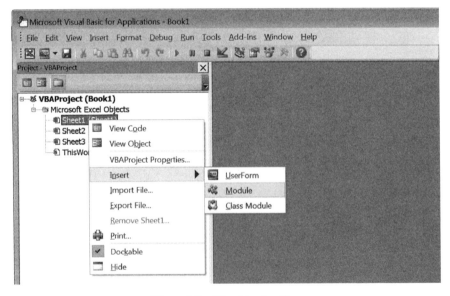

Figure 1.5   Visual basic editor.

4. Execute the program: in the VBE, select the module, and click the "Play" ribbon or press F5. In the Excel worksheet (Fig. 1.3), click the [Macros] button, choose the macro to be run, and click [Run]. A command button for a specific macro can be inserted in the Excel worksheet to facilitate the execution. See Section 1.2.4 for details.

1.2.3 The Macro Recorder

The macro recorder can record the actions that you perform in the Excel worksheet, such as building a chart or typing words, and transfer the actions into the macros in the VBE. This will be useful when you do not know how to code the actions and need to repeat them later. However, the macro recorder cannot handle codes that involve using the *For* loop or other repetitive loops and assigning variables. Different environments in Excel may generate different codes for the same task. Nevertheless, it can be a handy tool for learning new VBA codes. To record a macro, do the following:

1. Open the macro recorder: in the developer mode (Fig. 1.3), click [Record Macro].
2. Type the name to be used for the macro and a description of it so that you can recognize the macro next time (Fig. 1.6), then click [OK]. Note that the name should begin with a letter and contain no spaces or special characters.
3. Perform the tasks to be recorded; for example, type "Hello World" in cell A1.
4. Stop the macro recorder: click the [Stop Recording] button.
5. Go to the VBE to see the codes generated by the computer (Fig. 1.7).

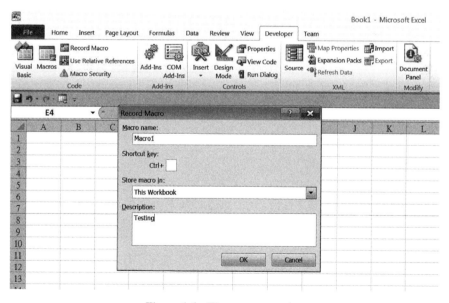

Figure 1.6 The macro recorder.

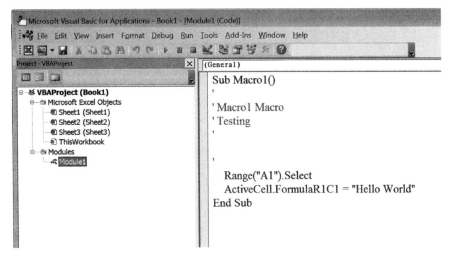

Figure 1.7   The recorded codes.

The recorder creates a new *Sub* module in the VBE (Fig. 1.7). To run this macro, just click [Macro] at the top ribbon in Excel and select the macro you want to run. In the recorded codes, the words following the symbol ' are not executed and serve only as comments. These comments are added to the codes to increase the readability for other users. It is a good programming habit to provide comments inside the codes to explain the details of the algorithm or define the variables. Comments can also be added by putting the keyword *Rem* at the beginning of the line.

1.2.4   Setting Up a Command Button

To run a specific macro in the Excel spreadsheet without selecting the macro procedure list, it is more convenient to designate a command button for each frequently used macro. To run the macro, the user just needs to press the command button. To insert a command button, follow the following procedure:

1. Click the [Insert] icon in the developer mode ribbon, and click the Command Button under [Form Controls] (Fig. 1.8).
2. Drag the mouse over a rectangle in the spreadsheet and release, then select the macro for the button.
3. To edit the button, left-click the name of the command button to change the name. Right-click the command button and select [Assign Macros] (Fig. 1.9) to change the macro.
4. Click on the command button to run the macro.

With this command button, users can quickly execute a macro.

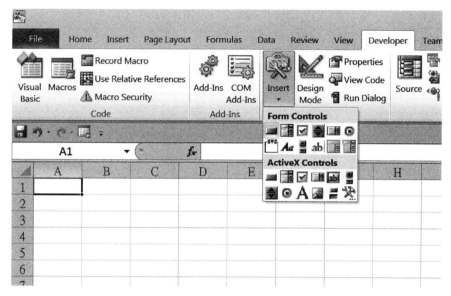

Figure 1.8　Creating command button.

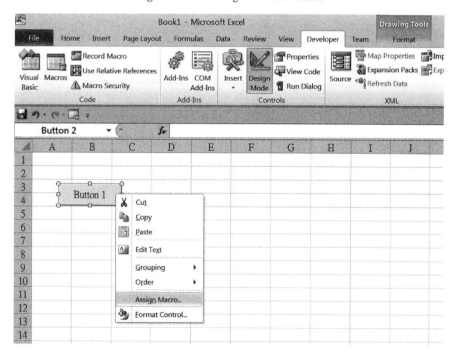

Figure 1.9　Assigning a macro to a command button.

1.3   VBA PROGRAMMING FUNDAMENTALS

1.3.1   Declaration of Variables

A variable in programming is the name of a place in the computer's memory where some values or objects can be stored. To declare a variable in VBA, we use the following statement:

```
Dim varname [As vartype],
```

where *varname* is the variable name and *vartype* is the variable type. A variable name must begin with a letter and contain only numeric and letter characters and underscores. The name should not be the same as a VBA reserved word, such as *Sub, Function, End, For, Optional, New, Next, Nothing, Integer,* and *String*. However, VBA does not distinguish between cases.

For the [*As vartype*] part, it is optional to specify the type of variable. This is different from other programming languages, which require the programmer to explicitly define the data type of each variable used. However, if you do not specify the data type explicitly, VBA will be slower to execute and use memory less efficiently.

1.3.2   Types of Variables

Every variable can be classified into one of four basic types: string data, date data, numeric data, and variant data. The string data type is used to store a sequence of characters. The date data type can hold dates and times separately and simultaneously. The types used most frequently in this book are numeric data and variant data.

There are several numeric data types in VBA, and the details of each type are listed in Table 1.1. In general, it is more efficient to use the data type that uses the smallest number of bytes. This can significantly reduce the computational time for simulations.

The variant data type is the most flexible because it can store both numeric and non-numeric values. VBA will try to convert a variant variable to the data type that can hold the input data. Defining [*As vartype*] is optional part, so an undeclared type of variable will be stored as *Variant* by default.

A variant type variable can also hold three special types of value: error code, *Empty* (indicating that the variable is empty and is not equal to 0, *False*, an empty string, or other value), and *Null* (the variable has not been assigned to memory and is not equal to 0, *False*, an empty string, *Empty*, or other value).

The following codes show some examples of variable declaration statements:

```
Dim x As integer
Dim z As string
z = "This is a string"
Dim Today As Date
Today = #1/9/2014# 'defined using month/day/year format
```

TABLE 1.1 Numeric Data Type

Type	Short Hand	Range	Description
Byte		0 to 255	Unsigned, integer number
Boolean		True(-1) or False(0)	Truth value
Integer	%	$-32,768$ to $32,767$	Signed integer number
Long	&	$-2,147,483,648$ to $2,147,483,647$	Signed integer number
Single	!	$\pm 3.402823E38$ to $\pm 1.401298E-45$	Signed single-precision floating-point number
Double	#	$\pm 1.79769313486231E308$ to $\pm 4.94065645841247E-324$	Signed double-precision floating-point number
Decimal		$\pm 7.922819251426433759E28$ with no decimal point and $\pm 7.9228162514264337 59354$ with 28 digits behind the decimal point	Cannot be directly declared in VBA; requires the use of a Variant data type

1.3.3 Declaration of Multivariable

We use the following statement to declare several variables:

```
Dim x As Integer, y As Integer, z As Integer
```

However, the declaration that

```
Dim x, y, z As Integer
```

denotes z as the Integer type only, while x and y are declared as variant types. We can use shorthand (Table 1.1) to improve the cleanness and readability of the program:

```
Dim x#, y#, z As Double
```

1.3.4 Declaration of Constants

Constants are declared in a *Const* Statement as follows:

```
Const interest_rate as Double = 0.02
Const dividend_yield = 0.02 'without declaring the constant type
Const option_type as String = "Put"
```

1.3.5 Operators

This section introduces the assignment operator, mathematical operators, comparative operators, and logical operators. The equal sign (=) is an assignment operator that

TABLE 1.2   VBA Logical Operators

Operator	Uses
Not	Performs a logical negation on an expression
And	Performs a logical conjunction on two expressions
Or	Performs a logical disjunction on two expressions
Xor	Performs a logical exclusion on two expressions
Eqv	Performs a logical equivalence on two expressions
Imp	Performs a logical implication on two expressions

is used to assign the value of an expression to a variable or constant. An expression is a combination of keywords, operators, variables, and constants that yields a string, number, or object.

For example,

```
y = 3 * 2
y = y * 6
```

Then y is evaluated as 36.

Other common mathematical operators include *addition* (+), *multiplication* (∗), *division* (/), *subtraction* (−), and *exponentiation* (^).

VBA also supports the same comparative operators used in Excel formulas: equal to (=), greater than (>), less than (<), greater than or equal to (>=), less than or equal to (<=), and not equal to (<>).

Table 1.2 lists the logical operators and their functions in VBA.

1.3.6   User-Defined Data Types

VBA provides the *Type* statement to allow users to create a more complex custom data type or user-defined data types (UDTs). The syntax for creating a UDT is as follows:

```
[Private | Public] Type typename
    [element_name As vartype]
    [element_name As vartype]
    ...
End Type
```

[*Private|Public*]: (optional) this is *Public* by default. If it is declared as *Private*, the UDT can only be declared in the same module as the UDT.

typename: (required) this is the name of the UDT, and it follows the standard variable naming conventions.

element_name: (required) this is the name of the elements within a UDT, and it follows the standard variable naming conventions.

vartype: (required) unlike declaring ordinary variables, the elements within a UDT must be assigned a data type, which can be any variable type (including *Variant*) or a UDT.

UDT can be defined at the top of the module before any procedures. To refer to the subelements within the UDT, use the period (.) operator. See the following example for illustration.

Example 1.1   *The following code defines a nested UDT, which stores the name and coordinates of a point.*

```
Type Coordinate
    x As Double
    y As Double
End Type

Type Point
    name As String
    z As Coordinate
End Type

Sub UDTEx1()
    'Declare p1 as UDT Point
    Dim p1 as Point

    'Assigning the values
    p1.name = "A"
    p1.z.x = 3.5
    p1.z.y = 3.1

    'Print out the values to spreadsheet
    Cells(1, 1) = p1.name
    Cells(2, 1) = p1.z.x
    Cells(3, 1) = p1.z.y
End Sub
```

1.3.7   Arrays and Matrices

An array is a collection of variables of the same type that have a common name. The index numbering makes it easy for users to perform looping in repetitive tasks.

The following statement declares a one-dimensional (1D) array:

```
Dim varname(LowerIndex to UpperIndex) As vartype.
```

In this way, users can access the variables with *varname*(LowerIndex), *varname*(LowerIndex +1), ..., *varname*(UpperIndex).

If only the upper index is specified, that is,

```
Dim varname(UpperIndex) As vartype ,
```

VBA will assume that 0 is the lower index.

A multidimensional array can be declared as:

```
Dim varname(LowerIndex1 to UpperIndex1, LowerIndex2 to _
UpperIndex2,...,LowerIndexN to UpperIndexN) As vartype .
```

Note that both the lower index and the upper index must be a constant or a number. A *dynamic array* should be used for the variable index, which does not have a preset number of elements. The following statement declares a dynamic array.

```
Dim varname() As vartype
```

Before the dynamic array is used, a *ReDim* statement should be inserted to specify the number of elements in the array. For example,

```
ReDim varname(LowerIndex to UpperIndex) .
```

To declare a matrix of size $m \times n$ containing real numbers, use the following statement.

```
Dim matrixmn() As Double
ReDim matrixmn(1 To m, 1 To n)
```

1.3.8   Data Input and Output

One advantage of using Excel VBA is that it can link the VBE and worksheet so that users can read in and print out data in the worksheet and execute the programs written in VBE. The following statements are usually used for input and output data, respectively.

```
'Read in data
Var = Cells(i, j)

'Print out data
Cells(i, j) = Var ,
```

where i and j denote the row number and the column number of a cell, respectively.

1.3.9   Conditional Statements

Conditional statements allow users to perform different tasks subject to different conditions. The two main conditional statements in VBA are *If-then-else* and *Select-Case* statements. There are two forms of the *If-then-else* statement: single-lined and multi-lined. Only one statement is allowed in the single-lined form, whereas many statements can be inserted in the multi-lined form. The syntax of the *If-then-else* statements is as follows:

```
'the Else clause is optional
If [condition] Then [statement] (Else [elseStatement])
'... represents other more statements can be included
'these Else clauses are also optional
If [condition] Then
    [statement]
    ...
ElseIf [elseif condition1] Then
    [Statement]
    ...
ElseIf [elseif condition2] Then
    [Statement]
    ...
Else
    [Statement]
    ...
End If
```

In the conditional part of the statement, the user needs to specify an expression that can be evaluated as True or False. The comparative operators and logical operators in Table 1.2 can help to express more complex conditions.

The *Select-Case* statement is useful for choosing among three or more options and is a good alternative to the *If-Then-Else* statement. The syntax for *Select-Case* is as follows:

```
Select Case [testexpression]
  Case expressionlist-n
     [instructions-n]
     ...
  Case expressionlist-n
     [instructions-n]
     ...
  Case Else
     [default_instructions]
     ...
End Select
```

1.3.10   Loops

The use of the loops algorithm allows users to perform certain tasks several times. *For-Next* loops and *Do* loops are widely used in VBA programming. In particular, *For-Next* loops are frequently used in simulations. The syntax for a *For-Next* loop is as follows:

```
For counter = startValue To endValue [Step nStep]
    [statements]
    [Exit For]
    [statements]
Next counter
```

If the *Step nStep* part is omitted, the counter will increase by 1 each time. We can set *nStep* to be *n*, and the counter will then increase by *n* each time.

For a *Do Loop*, the syntax is as follows:

```
Do [do_condition]
    [statements]
    [Exit Do]
    [statements]
Loop [loop_condition]
```

Although both the *do_condition* and the *loop_condition* are optional, only one of them can be used for a *Do Loop*. If both are omitted, then the user must specify a condition and call `Exit Do` to end the loop. Otherwise, the program will not terminate. The syntax is the same for *do_condition* and *loop_condition*.

```
While|Until condition
```

For *While*, the loop will continue as long as *condition* is *True*. For *Until*, the loop breaks once *condition* becomes *True*. Whether to use *While* or *Until* depends solely on the programmer's preference, as the same task can be performed by either loop. However, whether to put the condition after *Do* or *Loop* depends on the situation, because if it is put after *Loop*, then the loop is repeated at least once. The following example illustrates the uses of different loops to perform the same task.

Example 1.2   *Use five different methods to print out 1 to 10 to cells A1 to A10.*

```
'For Loop
For i = 1 to 10
    Cells(i, 1) = i
Next i

'Do Loop Method 1
i = 1
Do while i <= 10
   Cells(i, 1) = i
   i = i + 1
Loop

'Do Loop Method 2
i = 1
Do Until i > 10
    Cells(i, 1) = i
    i = i + 1
Loop

'Do Loop Method 3
i = 1
Do
    Cells(i , 1) = i
```

```
    i = i + 1
Loop while i <= 10

'Do Loop Method 4
i = 1
Do
    Cells(i, 1) = i
    i = i + 1
Loop until i > 10
```

1.3.11   Sub Procedures and Function Procedures

Large programs often need to be divided into smaller pieces for easier management and maintenance. In VBA, a *procedure* is basically a set of computer codes that performs certain tasks. There are two types of procedures: a *Sub* procedure and a *Function* procedure. A *Sub* procedure performs tasks but does not return values, while a *Function* procedure returns a value at the end of the procedure.

The syntax that defines a *Sub* procedure is as follows:

```
[Private|Public] [Static] Sub name ([arglist])
    [statements]
End Sub
```

  Private|Public: (optional) the *Sub* is *Public* by default if *Public* or *Private* is omit-
      ted. *Public* indicates that the *Sub* is accessible by other *Sub*s or *Function*s in
      all modules. *Private* indicates that the *Sub* is accessible only to the *Sub*s and
      *Function*s in the same modules.

  Static: (optional) *static* indicates that all local variables in the *Sub* are preserved
      at the end of the *Sub*. If *Static* is omitted, the values of the local variables will
      be reset each time the *Sub* ends.

  name: (required) this is the identifier of the *Sub*. It follows the standard variable
      naming conventions and must be unique and cannot be the same as the identifier
      of other *Sub*s, *Function*s, classes etc.

  arglist: (optional) this is a list of variables representing the parameters that are
      passed to the sub when it is called. Multiple variables are separated by commas.
      If the procedure uses no arguments, a set of empty parentheses is required.

  statements: (optional) this refers to any group of statements to be executed within
      the *Sub*.

Example 1.3   *The following procedure, SubEx2, calculates var1 +var2 and outputs
the result in cell A1:*

```
Sub SubEx2(var1, var2)
    Cells(1, 1) = var1 + var2
End Sub
```

To call the *Sub*, use one of the following two statements where x, y can also be replaced by other constants or variables.

```
Call SubEx2(x, y)
SubEx2 x, y
```

Instead of just specifying the name of the parameters, each parameter in *arglist* can be specified with the following syntax:

```
[Optional] [ByRef|ByVal] varname [As vartype] [= defaultvalue]
```

Optional: (optional) this indicates that this parameter is optional and will take the *defaultvalue* as its value if it is omitted when the *Sub* is called.

ByRef|ByVal: (optional) the parameter is passed *ByRef* by default. *ByRef* and *ByVal* indicate whether the parameter is passed by value or by address. When calling with *ByRef*, the memory address of the parameter is passed to the procedure and any change in the parameter value in the procedure will change the original parameter. For *ByVal*, a copy of the value of the parameter is passed so the original parameter will not be affected.

varname: (required) this is the identifier of the parameters.

vartype: (optional) the variable type is *Variant* by default. It is the variable type of the parameter passed, which can be any of the variable types or a UDT. If the variable that is passed when calling the *Sub* does not match, an error "ByRef/ByVal argument type mismatch" is shown.

defaultvalue: (optional) this is the value that the parameter will take when the parameter is not specified and the *Sub* is called.

Example 1.4   *The following codes demonstrate the difference between ByRef and ByVal:*

```
Sub SubEx3_Run()
    Dim x as integer, y as integer
    x = 1
    y = 1
    Call SubEx3(x, y)
    Cells(1, 1) = x
    Cells(2, 1) = y
End Sub

Sub SubEx3(ByRef var1 as integer, ByVal var2 as integer)
    var1 = var1 + 1
    var2 = var2 + 1
End Sub
```

Cell A1 shows that 2, as the change in the value of *var1* in *SubEx3*, actually changes the value of *x*. Cell A2 shows that 1, as the change of the value of *var2* in *SubEx3*, does not affect the value of *y*.

VBA also allows the user to create a *Sub* that takes an arbitrary number of parameters using *ParamArray*. When using *ParamArray*, the parameters can be passed only by reference and declared as the *Variant* type. They are stored in an array with the parameter's name. To declare such a *Sub*, use

```
Sub SubEx4(ParamArray var())
    [statements]
End Sub
```

Unlike a *Sub* module, a *Function* can be used in an Excel spreadsheet as a user-defined function. The syntax that defines a *Function* is as follows:

```
[Private|Public] [Static] Function name ([arglist, ...]) [as vartype]
    [statements]
End Sub
```

For *Private|Public*, *Static*, *name*, and *arglist*, a *Function* is identical to a *Sub*. The only difference between the declaration of *Function* and *Sub* is that when defining *Function*, the user may want to define the return type *vartype*. Otherwise, the return type is *Variant* by default. To return a value for a *Function*, the user just needs to store that value in a variable with the same name as the function name. To call a *Function*, use one of the following statements:

```
Call FuncName(x, y)
FuncName x, y
z = FuncName(x, y)
```

Note that the first two statements are identical to those used for *Sub*, so one can treat *Function* as *Sub* if the return value does not matter. For the third statement, the return value will be stored in *z*.

As *Sub* cannot return a value, to accomplish certain tasks, it may be necessary to use global variables or pass the variables by reference. Example 1.5 calculates *var1* + *var2* and outputs the result into cell A1, which is analogous to Example 1.3 using *Function*.

Example 1.5   *The following code calculates* 3 + 4 *by calling Function FuncEx4 and outputs the sum of the two numbers, 5, into cell A1.*

```
Sub SubEx4()
    Cells(1, 1) = FuncEx4(3, 4)
End Sub

Function FuncEx4(var1 as integer, var2 as integer) as integer
    FuncEx4 = var1 + var2
End Function
```

TABLE 1.3 Common Built-In Math Functions in VBA

Function	Return Value	Math Expression
Abs(x)	Absolute value of the x	$|x|$
Atn(x)	Arc-tangent of x in radians	$\tan^{-1} x$
Cos(x)	Cosine of x	$\cos x$
Exp(x)	Exponential of x	e^x
Int(x)	The integral part of x	$[x]$
Log(x)	Natural logarithm of x,	$\ln x$
Round(x[, dp])	x rounded to dp decimal place dp is 0 by default if omitted	
Sgn(x)	Number indicates the sign of x -1 for $x < 0$, 0 for $x = 0$, 1 for $x > 0$	$|x|/x$
Sin(x)	Sine of x	$\sin x$
Sqr(x)	Square root of x	$\sqrt{x}$
Tan(x)	Tangent of x	$\tan x$

1.3.12 VBA's Built-In Functions

VBA has a variety of built-in functions that can simplify calculations and operations. For a complete list of functions, please refer to the VBA Help System. In the VBE, you can type "VBA" to display a list of VBA functions. Table 1.3 shows some of the commonly used built-in mathematical functions and their return values in descriptive and mathematical forms.

Remarks: If the input number is negative, then the function *Int* returns the first negative integer that is less than or equal to the number and the *Fix* function returns the first negative integer greater than or equal to the number. For example, *Int*(-8.3) returns -9, whereas *Fix*(-8.3) gives -8.

Excel VBA also allows users to use Excel worksheet functions such as *Average*, *Stdev*. To call the worksheet functions, use one of the following commands:

```
Application.FunctionName([arglist])
WorksheetFunction.FunctionName([arglist])
Application.WorksheetFunction.FunctionName([arglist])
```

For example, to calculate $\sin^{-1} 0.5$, which is not provided in VBA's built-in function library but is included in Excel, one can use

```
x = Application.Asin(0.5).
```

This returns the value 0.5236 ($\approx \pi/6$) and is stored in x. Note that not all Excel worksheet functions can be used in VBA. In particular, worksheet functions that have an equivalent VBA function, such as sqrt and sin, cannot be used. For a complete list of Excel worksheet functions, please refer to the Excel help pages.

2

BASIC PROPERTIES OF FUTURES AND OPTIONS

2.1   INTRODUCTION

A financial derivative is a security whose value depends on the values of other more elementary securities, such as equities, bonds, and commodities. Forward contracts and futures contracts are two typical derivatives trading in the financial market. The primary use of forward and futures contracts is to hedge against portfolio risk, but they also offer speculative opportunities to investors. Before introducing the properties of these contracts, we present some fundamental concepts in derivative pricing.

2.1.1   Arbitrage and Hedging

An arbitrage opportunity is a situation whereby an investor is able to enter into a trade, usually involving two or more markets, in which he/she can lock in a position with a positive probability of profit and a zero probability of loss. An arbitrage opportunity usually lasts for a very short time in an efficient market. In pricing derivatives, we want to make sure that the fair prices of the derivatives will not lead to any arbitrage opportunities.

As mentioned, forwards and futures are used to hedge against risk, which means they can be used to transfer the risk of unfavorable price fluctuations to other market participants. For example, assume that you are holding a share of a stock currently worth $45, and you have a deal with a counterparty that you will exchange that share with him for $50 one month later. One month later, you are sure to get $50 if your

Simulation Techniques in Financial Risk Management, Second Edition. Ngai Hang Chan and Hoi Ying Wong.
© 2015 John Wiley & Sons, Inc. Published 2015 by John Wiley & Sons, Inc.

counterparty honors the deal. In this way, you hedge the market risk of the stock price for a fixed return. The existence of derivatives markets facilitates hedging and also possible speculation with large leveraging.

Another important concept is risk-neutral pricing, which states that the price of derivatives determined as "risk-neutral" totally agrees with the price obtained in the real world. In the risk-neutral world, every security generates the same expected rate of return, which is the risk-free interest rate. An investor can only earn excessive returns because of "pure luck." Modern derivative pricing theory argues that no arbitrage is associated with the existence of a risk-neutral world for the valuation of derivatives.

2.1.2 Forward Contracts

A forward contract is usually an over-the-counter (OTC) agreement between the buyer and the seller, whereby the buyer agrees to buy an asset (long position) from the seller (short position) at a certain future time (maturity) for a prespecified price (delivery price). The contract is usually traded between two financial institutions or between a financial institution and one of its corporate clients, but it is not traded on an exchange.

At the time of initiation of the contract, the delivery price is chosen so that the value of holding the forward contract is zero for both parties. At maturity, the holder of the short position delivers the asset to the holder of the long position in return for a cash amount equal to the delivery price. At the time the contract is entered into, the delivery price equals the forward price. As time passes, the delivery price is fixed, but the new forward price for the same underlying asset with the same maturity changes from time to time. These forward prices make the contract zero value at each time point. Therefore, the forward price generally does not equal the delivery price except at the beginning of the contract.

In the following, we determine the fair price of a forward contract. Let S_t be the price of the underlying asset at current time t, K be the delivery price, T be the maturity time of the contract, F_t be the forward price at time t, f_t be the value of the forward contract at time t, and r be the continuously compounded risk-free interest rate, which is assumed to be a constant. For simplicity, we assume there is no transaction cost in the market, the borrowing and lending rate are the same, and the trading profits have the same tax rate. At the initial time $t = 0$, the forward price equals the delivery price:

$$F_0 = K \quad \text{and} \quad f_0 = 0.$$

For a continuously compounding interest rate r, a zero-coupon bond paying \$1 at future time T is worth $e^{-r(T-t)}$ at time $t \leq T$. To determine the forward price, we construct two portfolios with the same payoff at maturity T under all scenarios. Then, these two portfolios should have the same price at current time t. This concept is referred to as the law of one price. No arbitrage implies that the prices of the two portfolios must be the same. We consider two cases of the underlying asset.

1. No intermediate income from the underlying asset.
   This kind of asset includes non-dividend-paying stocks and zero coupon bonds. Consider two portfolios at time t:

   A. Long a forward contract with delivery price K and invest $Ke^{-r(T-t)}$ in the bank for a risk-free interest rate r.

   B. Long one unit of the underlying asset S_t.

   Both portfolios will pay the holder one unit of the asset at time T; therefore, their current prices should be the same. Otherwise, investors can always long the cheaper portfolio and short the other one to gain a risk-less profit at maturity T. Hence, if there is no arbitrage, then

$$f_t + Ke^{-r(T-t)} = S_t.$$

The forward price F_t is the delivery price such that the forward contract has zero value at current time t. Therefore, we have

$$0 + F_t e^{-r(T-t)} = S_t,$$
$$\Rightarrow \quad F_t = S_t e^{r(T-t)}. \tag{2.1}$$

Example 2.1   *Consider a 6-month forward contract on a stock worth $13.50 per share at maturity. Assuming the current stock price is $12.00, the risk-free rate is 5.25%, and there is no dividend in the next 6 months. The forward price can be determined as*

$$F_0 = 12e^{0.0525(0.5-0)}$$
$$= 12.32.$$

If the forward price is cheaper than the delivery price, it is possible to obtain arbitrage by shorting the forward contract and borrowing $12 from the bank at a rate of 5.25% to buy one share of the stock now. The investor does not need to put any money in this portfolio at its initiation, and this is called a self-financing portfolio. Six months later, the investor can deliver the share of stock for $13.50 and pay back $12.32 to the bank. He will be sure to earn $13.50-$12.32=$1.18 after 6 months.

2. With a known cash income.
   Let I be the present value of the income to be received from the underlying asset during the life of the forward. Again, we construct two portfolios as follows:

   C. Long a forward contract with delivery price K and invest $Ke^{-r(T-t)}$ in the bank for a risk-free interest rate r.

   D. Long one unit of the underlying asset S_t and borrow an amount, I, from the bank at the risk-free interest rate r.

In portfolio D, the cash income I from holding the stock is paid back to the bank. These two portfolios give both holders a share of the stock at maturity; thus their current prices must be the same to avoid arbitrage. We have

$$f_t + Ke^{-r(T-t)} = S_t - I.$$

The forward price is the delivery price that ensures the forward has a zero value at the time of initiation. Therefore,

$$0 + F_t e^{-r(T-t)} = S_t - I,$$
$$\Rightarrow \qquad F_t = (S_t - I)e^{r(T-t)}. \qquad (2.2)$$

Example 2.2   *Consider a 6-month forward contract on a stock worth \$11.50 per share at maturity. Assuming the current stock price is \$13.00 and the risk-free rate for 6-month maturity is 5.25%, there will be a \$1.20 dividend to be paid 3 months from now. The 3-month interest rate is 5.1%. The forward price can be determined as*

$$F_0 = \left(13 - 1.2e^{-0.051(0.25-0)}\right) e^{0.0525(0.5-0)}$$
$$= 12.13.$$

If the investor longs one unit of the forward contract, shorts one share of the stock, and invests \$$(13 - 1.2e^{-0.051(0.25)})$ in the bank at a rate of 5.25% for 6 months and $\$1.2e^{-0.051(0.25)}$ at a rate of 5.1% for 3 months, he will have a risk-less profit after 6 months. This is a self-finance strategy. After 3 months, the 3-month deposit of \$1.20 is paid out as a dividend. At maturity, the investor can have one share of stock for \$11.50 and re-pay the loaned stock. Therefore, he will gain \$12.13-\$11.50=\$0.63 without any risk.

If the dividend is paid out continuously at an annual rate q, then q is called the dividend yield and the forward price can be determined using similar arguments. Specifically, we keep the portfolio A and revise the portfolio B to long $e^{-q(T-t)}$ units of the stock, so that we will have exactly one share of the stock at maturity. The present value of holding the forward contract is

$$f_t = S_t e^{-q(T-t)} - Ke^{-r(T-t)}.$$

The forward price is given such that the value of the forward equals zero, so we have

$$F_t = S_t e^{(r-q)(T-t)}. \qquad (2.3)$$

Example 2.3   *Consider a 6-month forward contract on a stock worth $14 per share at maturity. Assuming the current stock price is $13.40 and the risk-free rate for 6-month maturity is 5.2%, the dividend yield is 2%. The forward price can be determined as*

$$F_0 = 13.4e^{(0.052-0.02)(0.5-0)}$$

$$= 13.62.$$

If the investor shorts one unit of the forward contract, borrows $(13e^{-0.02(0.5)})$ from the bank at a rate of 5.2% to buy $e^{-0.02(0.5)}$ shares of stock with the dividends being reinvested in the stock after 6 months, he will have a risk-less profit of $\left(14 - 13.4e^{(0.052-0.02)(0.5)}\right) = \$0.38.$

The value of the forward contract changes with time. For example, assume that you are holding a 6-month forward contract with a delivery price of $10 for a share of the stock. However, after 3 months, suppose the delivery price of a new 3-month forward contract is $12, then your original forward contract is now worth $(12 - 10)e^{-r(0.5-0.25)}$ dollars because you can short a new contract for $12 and your position will be closed out 3 months later. In general, the value of a forward contract is given by the formula

$$f_t = (F_t - F_0)e^{-r(T-t)}. \tag{2.4}$$

2.1.3   Futures Contracts

A futures contract is an agreement between two parties to buy or sell an asset at a certain time at the future price. Unlike forward contracts, futures are normally traded on an exchange, and this can eliminate the default risk of the counterparty. However, the values of futures contracts are also marked-to-market, meaning that the values are determined each day according to the market price. Therefore, investors in futures can be subject to a margin call.

The exact delivery date of futures is not usually specified, in contrast to forward contracts. A futures contract is referred to by its delivery month, and the exchange center specifies the period during the month when the delivery must be made. Nowadays, a lot of futures are settled by cash instead of actual delivery of the assets. When the interest rate is constant (even deterministic), the theoretical prices of forward and futures contracts with the same delivery date are the same. To show this, we denote F_t as the futures price and $\tilde{F}_t$ as the forward price. Now consider two different trading strategies with futures and forward contracts, respectively, as follows:

Strategy A. Long e^r units of futures on day 1, close out the futures on day 2 and long e^{2r} units of futures on day 2, close out on day 3, and so on, according to Table 2.1, and invest F_0 in a risk-free asset.

Strategy B. Long e^{rT} units of forward contracts with the forward price $\tilde{F}_0$ and invest $\tilde{F}_0$ in a risk-free asset.

TABLE 2.1   Strategy A for Longing Futures Contracts

Day	0	1	2	$\cdots$	$T-1$	T
Future price	F_0	F_1	F_2	$\cdots$	F_{T-1}	F_T
Future positions	e^r	e^{2r}	e^{3r}	$\cdots$	e^{Tr}	0
Gain/loss	0	$(F_1 - F_0)e^r$	$(F_2 - F_1)e^{2r}$	$\cdots$	$\cdots$	$(F_T - F_{T-1})e^{rT}$
Gain/loss at time T	0	$(F_1 - F_0)e^{rT}$	$(F_2 - F_1)e^{rT}$	$\cdots$	$\cdots$	$(F_T - F_{T-1})e^{rT}$
Total gain/loss at time T	$0 + (F_1 - F_0)e^{rT} + (F_2 - F_1)e^{rT} + \cdots + (F_T - F_{T-1})e^{rT}$ $= (F_T - F_0)e^{rT} = (S_T - F_0)e^{rT}$					

At maturity date T, the payoff of strategy A is $(S_T - F_0)e^{rT} + F_0 e^{rT} = S_T e^{rT}$, whereas the payoff of strategy B is $(S_T - \tilde{F}_0)e^{rT} + \tilde{F}_0 e^{rT} = S_T e^{rT}$. According to the no-arbitrage argument, these two strategies should yield the same value at any moment before time T. Therefore, we have

$$F_0 = \tilde{F}_0. \tag{2.5}$$

The variance of a portfolio can represent the risk level it exposes the investor to. Suppose that you need to sell N_A units of stock in the future, at time t, how many units of futures should you short now so that the variance of your portfolio is minimized? To answer this question, let N_F be the units of futures you should short, with a hedge ratio of $h = N_F/N_A$. This is called a static hedge, as the hedge is carried out only once at time 0 and will not need to be adjusted later. In contrast, a dynamic hedge requires continuous re-balancing of the portfolio weights. More examples of dynamic hedging are introduced in later chapters. The payoff Y_t of the futures portfolio at maturity is given by

$$
\begin{aligned}
Y_t &= N_A S_t - N_F (F_t - F_0)e^{-r(t-t)} \\
&= N_A S_t - N_F (F_t - F_0) \\
&= N_A S_0 - N_A (S_t - S_0) - N_F (F_t - F_0) \\
&= N_A S_0 - N_A \Delta S_t - N_F \Delta F_t \\
&= N_A S_0 - N_A (\Delta S_t - h \Delta F_t).
\end{aligned}
$$

Let σ_S^2 be the variance of the stock price, σ_F^2 be the variance of the futures price, and ρ be the correlation of the stock price and the futures price. The variance of the portfolio can be evaluated as:

$$
\begin{aligned}
\text{Var}(Y_t) &= N_A^2 \, \text{Var}(\Delta S_t - h \Delta F_t) \\
&= N_A^2 (\sigma_S^2 + h^2 \sigma_F^2 - 2h\rho \sigma_S \sigma_F).
\end{aligned}
$$

To minimize the variance with respect to the choice of h, we take the derivative of $Var(Y_t)$ with respect to h and set it to zero as follows:

$$\frac{d\,Var(Y_t)}{dh} = N_A^2(2h\sigma_F^2 - 2\rho\sigma_S\sigma_F) = 0$$

$$\Rightarrow \quad h^* = \rho\frac{\sigma_S}{\sigma_F}. \tag{2.6}$$

From the minimum-variance hedge ratio formula, we can also derive the hedging effectiveness as follows:

$$\frac{Var(\text{unhedged port.}) - Var(\text{hedged port.})}{Var(\text{unhedged port.})} = \frac{Var(N_A S_t) - Var(Y_t)}{Var(N_A S_t)}$$

$$= \frac{\sigma_S^2 - (1 - \rho^2)\sigma_S^2}{\sigma_S^2}$$

$$= \rho^2. \tag{2.7}$$

Example 2.4 *Suppose we have a set of data on the stock and futures prices, as shown in Table 2.2, we can calculate the optimum hedge ratio as*

$$h^* = (0.0928 \times 0.00262)/0.00313$$

$$= 0.786$$

If we want to sell, say, 50,000 units of the stock at time t, we can calculate N_F as

$$N_F^* = h^* \times N_A$$

$$= 0.786 \times 50,000$$

$$= 39,300.$$

Therefore, if there are 1,000 units per futures contract, the portfolio will have the minimum variance if we short approximately 39 futures contracts.

TABLE 2.2   Data on Stock and Futures Prices

Month	ΔF	ΔS
1	0.021	0.029
2	0.035	0.020
⋮	⋮	⋮
15	−0.027	−0.032
Mean	−0.013	0.0138
σ	0.00313	0.00262
ρ	0.928	

2.2   OPTIONS

Options were first traded on an organized exchange in 1973. Since then, the option markets have experienced a dramatic growth. There are two basic types of options, namely call and put options. A call (put) option gives the holder the right, but not the obligation, to buy (sell) the underlying asset for a prespecified price (strike price K) at some future time. In contrast to forward and futures contracts, options will be exercised only if exercising is favorable to the holders.

Options can be further divided into American or European types. American options can be exercised at any time up to maturity T, while European options can only be exercised on the maturity date T. As the option holder will not lose anything in the worst situation (the option is just not exercised), a premium has to be paid in exchange for this privilege. The premium that must be paid to the seller is the fair value of the option. Derivative pricing theory studies methods of finding fair premiums for different kinds of financial derivatives.

We can either long or short an option, so there are four kinds of payoffs in general, as summarized in Table 2.3. Figure 2.1 shows the graphs of the payoff functions. The payoff functions reveal some interesting properties of the options related to underlying asset. Let C_A be the American call price with maturity T and strike K, C_E be the corresponding European call, and P_A and P_E be the American put and European put, respectively. Some option properties can be derived from the following.

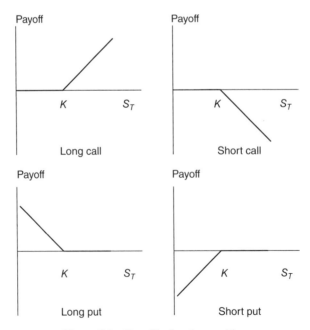

Figure 2.1   Payoffs of option positions.

TABLE 2.3 Payoffs of Different Options with Strike Price K

Type	Call	Put
Long	$\max(S_T - K, 0)$	$\max(K - S_T, 0)$
Short	$-\max(S_T - K, 0)$	$-\max(K - S_T, 0)$

1. Upper bounds.
   Whatever the price of an underlying asset, the value of a call option (with payoff $\max(S - K, 0)$) can never be worth more than the stock, and an American call is always worth more than its European counterpart because it can be exercised at any time, including at maturity, so we have

   $$C_E \le C_A \le S.$$

   For put options, no matter how low the stock price becomes, the put can never be worth more than the strike price:

   $$P_E \le K \quad \text{and} \quad P_A \le K.$$

   Furthermore, for a European put, the option will not be worth more than K at maturity, so the current value of the put cannot be larger than the present value of the strike:

   $$P_E \le Ke^{-r(T-t)}.$$

2. Lower bounds.
   The lower bounds for call options can be derived as follows:

   $$\max\left(S - Ke^{-r(T-t)}, 0\right) \le C_E \le C_A.$$

   To prove the aforementioned inequality, we consider two portfolios:
   A. Hold one unit of European call and K units of zero coupon bonds.
   B. Long one share of the stock.
   By comparing the stock price with the strike price on the maturity date, we can find the values of the two portfolios, as shown in Table 2.4.
   From the table, the value of portfolio A should be larger than that of portfolio B at any time. Otherwise, it is always possible to long portfolio A and short portfolio B to gain arbitrage. Together with the positive nature of options prices, we can deduce that

   $$S \le C_E + Ke^{-r(T-t)} \quad \text{and} \quad 0 \le C_E,$$
   $$\Rightarrow \qquad \max(S - Ke^{-r(T-t)}, 0) \le C_E.$$

TABLE 2.4   Payoffs of Portfolios A and B

	$S > K$	$S \leq K$
Portfolio A	S	K
Portfolio B	S	S

Similarly, we can obtain the inequality for put options:

$$\max(Ke^{-r(T-t)} - S, 0) \leq P_E \leq P_A.$$

This inequality can also be shown by applying the put-call parity to call options.

3. Put-call parity.

The prices of European call and put options with the same strike and maturity are related by the following put-call parity formula:

$$C_E + Ke^{-r(T-t)} = P_E + S. \tag{2.8}$$

To prove this by the no-arbitrage principle, we construct two portfolios:

C. Long a call option and K units of zero coupon bonds.

D. Long a put option and one share of stock.

At maturity, both portfolios have the same values (Table 2.5) regardless of the stock price, so these two portfolios should have the same present values according to the no-arbitrage principle. Note that the put-call parity relation only holds for European options. However, we can derive some inequality relations for American options. For non-dividend-paying assets, we have

$$P_A \geq P_E \qquad \text{and} \qquad C_A = C_E. \tag{2.9}$$

It is never optimal to early exercise a non-dividend-paying American call option, because doing so will gain $\max(S - K, 0)$, but the lower bound for the call option is

$$S - K < S - Ke^{-r(T-t)},$$

$$\Rightarrow \max(S - K, 0) \leq \max\left(S - Ke^{-r(T-t)}, 0\right) \leq C_A.$$

TABLE 2.5   Payoffs of Portfolios C and D

	$S > K$	$S \leq K$
Portfolio C	S	K
Portfolio D	S	K

Therefore, holding the call contract is actually worth more than exercising the contract. This also proves Equation 2.9. The following example provides some observations on the Hong Kong Hang Seng Index options market due to the put-call parity.

Example 2.5   *When the interest rate is very close to zero, the put-call parity relation gives*

$$C_E + K \approx P_E + S.$$

If there is no arbitrage opportunity in the market, the aforementioned condition needs to be satisfied. In the case where the underlying asset value is unknown to us, we can also check the relation by using call and put prices with the same maturity for two strike prices, K_1 and K_2. Let $C_E(K_1)$ be the call price with the strike price K_1 and denote a similar notation for $P_E(K_1)$; then we have

$$C_E(K_1) + K_1 \approx P_E(K_1) + S,$$
$$C_E(K_2) + K_2 \approx P_E(K_2) + S,$$
$$\Rightarrow \left(C_E(K_1) - C_E(K_2)\right) - \left(P_E(K_1) - P_E(K_2)\right) \approx K_2 - K_1.$$

To verify this claim, we check the prices of the Hang Seng Index options on September 12, 2014 (Fig. 2.2) with three different maturities. Both sides of the aforementioned formula are evaluated in the Excel worksheet. The results show that the market prices closely match the put-call parity relation.

4. Differences between American call and put prices.
   For a non-dividend-paying asset, we can further deduce the boundaries of the difference between the prices of American call and put options. According to the put-call parity,

$$P_A \geq P_E$$
$$= C_E + Ke^{-r(T-t)} - S$$
$$= C_A + Ke^{-r(T-t)} - S,$$
$$\Rightarrow \quad C_A - P_A \leq S - Ke^{-r(T-t)}.$$

To obtain the lower bound of the difference, consider the following two portfolios with the same maturity and same strike on the options:

E. Long a European call and hold K units of cash.

F. Long an American put and one unit of the asset.

A	B	C	D	E
Maturity=Nov,2014	Call Option		Put Option	
Strike K	24600	25200	24600	25200
Option Price	701	443	667	999
	[C(K1)-C(K2)]-[P(K1)-P(K2)]=			590
	K2-K1=			600
Maturity=Dec,2014	Call Option		Put Option	
Strike K	24200	25000	24200	25000
Option Price	1051	648	629	1027
	[C(K1)-C(K2)]-[P(K1)-P(K2)]=			801
	K2-K1=			800
Maturity=Jun,2015	Call Option		Put Option	
Strike K	26000	26600	26000	26600
Option Price	740	574	2512	2946
	[C(K1)-C(K2)]-[P(K1)-P(K2)]=			600
	K2-K1=			600

Figure 2.2   The prices of Hang Seng Index options on September 12, 2014.

When the American put is not exercised prematurely, portfolio F is worth $\max(K - S_T, 0) + S_T = \max(S_T, K)$ at maturity T. The value of portfolio E at maturity T is given by

$$\max(S_T - K, 0) + Ke^{r(T-t)} = \max(S_T, K) + K\left(e^{r(T-t)} - 1\right).$$

The value of portfolio E is larger than that of portfolio F if neither is exercised early. If the American put option is exercised prematurely at time $\tau < T$, the value of portfolio F at time τ is K while portfolio E is worth $C_E + Ke^{r(T-\tau)}$, which is greater than the value of portfolio F. The payoffs of these two portfolios are summarized in Table 2.6.

Portfolio E is worth more than portfolio F under all circumstances, so the present value of portfolio E should be larger than that of portfolio F:

$$C_E + K \geq P_A + S.$$

Note that $C_A = C_E$ for non-dividend-paying assets. By rearranging the aforementioned inequality, we obtain the boundaries as follows:

$$S - K \leq C_A - P_A \leq S - Ke^{-r(T-t)}. \tag{2.10}$$

TABLE 2.6 Payoffs of Portfolios E and F

	No Early Exercise	Early Exercise of American put at τ
	(Value at Maturity T)	(Value at Exercising Time τ)
Portfolio E	$\max(S_T, K) + K\left(e^{r(T-t)} - 1\right)$	$C_E + Ke^{r(T-\tau)}$
Portfolio F	$\max(S_T, K)$	K

TABLE 2.7 Properties of Stock Options

	C_E	P_E	C_A	P_A
Stock price	+	−	+	−
Strike	−	+	−	+
Maturity	+/−	+/−	+	+
Volatility	+	+	+	+
Risk-free rate	+	−	+	−
Dividends	−	+	−	+

The "+" sign indicates that the option is rising in value with an increase in the parameters; the "−" sign represents a decrease in value; and "+/−" represents an unclear influence on the price.

The aforementioned inequality also shows that when the interest rate $r \sim 0$, $S - K \approx C_A - P_A$. The put-call parity also implies that $C_A + K \approx P_E + S$ for non-dividend-paying assets. Therefore, we can deduce that

$$C_A - P_A \approx S - K \approx C_A - P_E$$
$$\Rightarrow \qquad P_A \approx P_E$$

for a near-zero interest rate. To price an option, a model of the stock price usually has to be specified except that the option can be perfectly replicated by other securities in the market. The next few chapters are devoted to the Black–Scholes model. Some qualitative properties related to the option parameters are summarized in Table 2.7.

2.3 EXERCISES

1. Assume today to be March 3, 2014, and the continuously compounding interest rate is 0.4% per annum. It is known that the interest rate will increase linearly over time to 1.2% until March 7, 2013. Consider a 1-year futures contract, a 1-year European call option, and a 1-year equity swap (ES) contract on a non-dividend-paying stock with a current price of $40. The ES with four transaction dates on June 2, 2014; September 2, 2014; December 2, 2014; and March

2, 2015 will deliver one unit of the underlying stock to the holder while receiving a constant amount of cash on each transaction date.

(a) What is the no-arbitrage forward price of a 1-year forward contract? Is it the same as the no-arbitrage futures price? Explain briefly.

(b) What is the ES price (the constant cash amount paid to the exchange for the stock on each transaction date)?

2. A 3-month European-style derivative with the following payoff is selling at $2:

$$\text{Payoff} = \min [\, \max(33 - S_T, 0), \max(S_T - 27, 0)].$$

At the moment, the underlying stock price is $30 and the European call prices with strikes, $27, $30, $33, are traded at $3.2193, $1.20, $0.2874, in order. Assume that the risk-free interest rate is 10% per annum for all maturities. What arbitrage opportunity does this create from this exotic derivative? Construct an arbitrage strategy in detail.

3. An out-range option has the same payoff as an ordinary option, except that it cannot be exercised if the terminal asset price falls within a predetermined range. A range-digital option pays the holder $1 if the terminal asset price falls into the prespecified range; otherwise, the holder receives nothing. Other things being equal, we introduce the following notations:

- Out-range put $= P_R(K, L, U)$, where K is the strike price and $[L, U]$ is the range.
- European put $= P_E(K)$, where K is the strike price.
- European call $= C_E(K)$, where K is the strike price.
- Range-digital option $= D(L, U)$, where $[L, U]$ is the range.

Suppose $K > U > L$. Show the Range-Digital-European (RDE) parity relation:

$$P_R(K, L, U) + (K - L)D(L, U) + (U - L)[e^{-rT} - D(0, U)]$$
$$= P_E(K) + C_E(L) - C_E(U).$$

4. A 4-month European call option on a dividend-paying stock is currently selling for $5. The stock price is $64, the strike price is $60, and a dividend of $0.80 is expected in 1 month. The risk-free interest rate is 12% per annum for all maturities. What opportunities are there for an arbitrageur?

5. Assume that the risk-free interest rate is 4% per annum with continuous compounding and that the dividend yield on a stock index varies throughout the year. In February, May, August, and November, the dividend yield is 6% per annum, and in other months it is 3% per annum. Suppose that the value of the index on July 31, 2010 is 300. What is the futures price for a contract that is deliverable on December 31, 2010?

6. A 1-year-long forward contract on a non-dividend-paying stock is entered into when the stock price is $40 and the risk-free interest rate is 5% per annum with continuous compounding.

(a) What are the forward price and the initial value of the forward contract?

(b) Six months later, the stock price is $45 and the risk-free interest rate is still 5%. What are the forward price and the forward value of the contract?

7. A company enters into a forward contract with a bank to sell a foreign currency for K_1 at time T_1. The exchange rate at time T_1 proves to be $S_1(> K_1)$. The company asks the bank if it can roll the contract forward under $T_2(> T_1)$ rather than settle at time T_1. The bank agrees to a new delivery price, K_2. Explain how K_2 should be calculated.

8. An ES is a contract that generalizes a forward contract. For a two-tenor ES, the long position will pay the short position K at each time point T_1 and T_2, where $T_1 < T_2$, while the short position will deliver one unit of the underlying asset S at both T_1 and T_2. Let $f(t, S)$ be the value of ES and F_t be the ES price, which makes the ES value zero at time $t < T_2$.

(a) What are the no-arbitrage pricing formulas for $f(t, S)$ and F_t?

(b) Consider that T_1 is 3 months from today and T_2 is 1 year from today. Suppose that the continuously compounded interest rate is a constant of 3%, and the underlying non-dividend-paying share is currently $10. What is the no-arbitrage value of K?

(c) What is the no-arbitrage price for an n-tenor ES? The n-tenor ES has n transaction dates at $T_1 < T_2 < \cdots < T_n$.

9. A minimum put option, P_{min}, gives the holder the right to sell the less expensive stock between S_1 and S_2 with a strike price of K on maturity.

(a) What is the payoff function of this option?

(b) Alternatively, a minimum call option, C_{min}, gives the holder the right to buy the less expensive stock between S_1 and S_2 with a strike price of K on the maturity date. Given that the payoff of an exchange option, C_X, is max$(S_2 - S_1, 0)$, use the no-arbitrage principle to show that

$$P_{min}(t, T) - C_{min}(t, T) = C_X(t, T) + Ke^{-r(T-t)} - S_2(t).$$

10. In Figure 2.2, we can see that the prices of the options on the Hang Seng Index that mature in June 2015 match the put-call parity relation, and the difference in the prices from the put-call relation for options that mature in December 2014 is small, which is reasonable due to the transaction cost. For the options that mature in November 2014, the discrepancy in the prices from the put-call parity is not small. What portfolio can you construct for an arbitrage opportunity?

11. Example 2.4 computes the minimum-variance hedge ratio for a specific data set. Now use the Hang Seng Index to compute the minimum-variance hedge ratio in an Excel worksheet, as in Table 2.2. Use the nearest 3-month daily mid-closing prices of the futures and the Hang Seng Index. The Excel functions *AVERAGE*, *STDEV*, and *CORREL* may be helpful in the computation.

The solutions and/or additional exercises are available online at http://www.sta.cuhk.edu.hk/Book/SRMS/.

3

INTRODUCTION TO SIMULATION

3.1 QUESTIONS

In this introductory chapter, we are faced with three basic questions as follows:

- What is simulation?
- Why does one need to learn simulation?
- What has simulation to do with risk management and, in particular, financial risk management?

3.2 SIMULATION

When faced with uncertainties, one tries to build a probability model. In other words, risks and uncertainties can be handled (managed) by means of stochastic models. However, in real life, building a full-blown stochastic model to account for every possible uncertainty is futile. One needs to compromise between choosing a model that is a realistic replica of the actual situation and choosing one whose mathematical (statistical) analysis is tractable.

However even equipped with the best insight and powerful mathematical knowledge, solving a model analytically is an exception rather than a rule. In most situations, one relies on an approximated model and learns about this model with approximated solutions. It is in this context that simulation comes into the picture.

Simulation Techniques in Financial Risk Management, Second Edition. Ngai Hang Chan and Hoi Ying Wong.
© 2015 John Wiley & Sons, Inc. Published 2015 by John Wiley & Sons, Inc.

Loosely speaking, one can think of simulations as computer experiments. It plays the role of the experimental part in physics. When one studies a physical phenomenon, one relies on physical theories and experimental verifications. When one tries to model a random phenomenon, one relies on building an approximated model (or an idealized model) and simulations (computer experiments).

Through simulations, one learns about different characteristics of the model, behaviors of the phenomenon, and features of the approximated solutions. Ultimately, simulations offer practitioners the ability to replicate the underlying scenario via computer experiments. It helps us to visualize the model, to study the model, and to improve the model.

In this book, we learn some of the features of simulations. We see that simulation is a powerful tool for analyzing complex situations. We also study different techniques in simulations and their applications in risk management.

3.3 EXAMPLES

Practical implementation of risk management methods usually requires substantial computations. The computational requirement comes from calculating summaries, such as value-at-risk, hedging ratio, market β, and so on. In other words, summarizing data in complex situations is a routine job for a risk manager, but the same can be said for a statistician. Therefore, many of the simulation techniques developed by statisticians for summarizing data are equally applicable in the risk management context. In this section, we study some typical examples.

3.3.1 Quadrature

Numerical integration, also known as quadrature, is probably one of the earliest techniques that requires simulation. Consider a one-dimensional integral

$$I = \int_a^b f(x)\,dx, \tag{3.1}$$

where f is a given function. Quadrature approximates I by calculating f at a number of points $x_1, x_2, \ldots, x_n$ and applying some formula to the resulting values $f(x_1), \ldots, f(x_n)$. The simplest form is a weighted average

$$\hat{I} = \sum_{i=1}^n w_i f(x_i),$$

where $w_1, \ldots, w_n$ are some given weights. Different quadrature rules are distinguished by using different sets of design points $x_1, \ldots, x_n$ and different sets of weights $w_1, \ldots, w_n$. As an example, the simplest quadrature rule divides the interval

$[a, b]$ into n equal parts, evaluates $f(x)$ at the midpoint of each subinterval, and then applies equal weights. In this case,

$$\hat{I} = \frac{b-a}{n} \sum_{i=1}^{n} f(a + (2i-1)(b-a)/(2n)).$$

This rule approximates the integral by the sum of the area of rectangles with base $(b-a)/n$ and height equal to the value of $f(x)$ at the midpoint of the base. For n large, we have a sum of many tiny rectangles whose area closely approximates I in exactly the same way that integrals are introduced in elementary calculus.

Why do we care about evaluating Equation 3.1? For one, we may want to calculate the expected value of a random quantity X with p.d.f. (probability distribution function) $f(x)$. In this case, we calculate

$$E(X) = \int xf(x)\,dx,$$

and quadrature techniques may become handy if this integral cannot be solved analytically. Improvements over the simple quadrature have been developed, for example, Simpson's rule and the Gaussian rule. We will not pursue the details in this case, but interested readers may consult Conte and de Boor (1980). Clearly, generalizing this idea to higher dimensions is highly nontrivial. Many of the numerical integration techniques break down for evaluating high dimensional integrals. (Why?)

3.3.2   Monte Carlo

Monte Carlo integration is a different approach to evaluating an integral of f. It evaluates $f(x)$ at *random* points. Suppose that a series of points $x_1, \ldots, x_n$ are drawn independently from the distribution with density $g(x)$. Now

$$I = \int f(x)\,dx = \int [f(x)/g(x)]g(x)\,dx = E_g\left\{\frac{f(x)}{g(x)}\right\}, \tag{3.2}$$

where E_g denotes expectation with respect to the distribution g. Now, the sample of points $x_1, \ldots, x_n$ drawn independently from g gives a sample of values $f(x_i)/g(x_i)$ of the function $f(x)/g(x)$. We estimate the integral (Eq. 3.2) by the sample mean

$$\hat{I} = \frac{1}{n} \sum_{i=1}^{n} \frac{f(x_i)}{g(x_i)}.$$

According to classical statistics, $\hat{I}$ is an unbiased estimate of I with variance

$$\text{Var}(\hat{I}) = \frac{1}{n} \text{Var}_g \frac{f(x)}{g(x)}.$$

As n increases, $\hat{I}$ becomes a more and more accurate estimate of I. The variance (verify) can be estimated by its sample version, namely,

$$\frac{1}{n^2} \sum_{i=1}^{n} \frac{f^2(x_i)}{g^2(x_i)} - \frac{\hat{I}^2}{n}. \tag{3.3}$$

Besides the Monte Carlo method, we should also mention that the idea of the quasi-Monte Carlo method has also enjoyed considerable attention recently. Further discussions on this method are beyond the scope of this book. Interested readers may consult the survey article by Hickernell, Lemieux, and Owen (2005).

3.4 STOCHASTIC SIMULATIONS

In risk management, one often encounters stochastic processes such as Brownian motions, geometric Brownian motion, and lognormal distributions. Although some of these entities may be understood analytically, quantities derived from them are often less tractable. For example, how can one evaluate integrals such as $\int_0^1 W(t)\, dW(t)$ numerically? More importantly, can we use simulation techniques to help us understand features and behaviors of geometric Brownian motions or lognormal distributions? To illustrate the idea, we begin with the lognormal distribution.

As the lognormal distribution plays such an important role in modeling the stock returns, we discuss some properties of the lognormal distribution in this section. Firstly, recall that if $X \sim N(\mu, \sigma^2)$, then the random variable $Y = e^X$ is lognormally distributed, that is, $\log Y = X$ is normally distributed with mean μ and variance σ^2. Thus, the distribution of Y is given by

$$G(y) = P(Y \leq y) = P(X \leq \log y)$$
$$= P((X - \mu)/\sigma \leq (\log y - \mu)/\sigma)$$
$$= \Phi((\log y - \mu)/\sigma),$$

where $\Phi(\cdot)$ denotes the distribution function of a standard normal random variable. Differentiating $G(y)$ with respect to y gives rise to the p.d.f. of Y. To calculate EY, we can integrate it directly with respect to the p.d.f. of Y or we can make use of the normal distribution properties of X. Recall that the moment-generating function of X is given by

$$M_X(t) = E(e^{tX}) = e^{\mu t + \frac{1}{2}\sigma^2 t^2}.$$

Thus,

$$EY = E(e^X) = M_X(1) = e^{\mu + \frac{1}{2}\sigma^2}.$$

By a similar argument, we can calculate the second moment of Y and deduce that

$$\mathrm{Var}(Y) = e^{2\mu + \sigma^2}(e^{\sigma^2} - 1).$$

To produce the densities of lognormal random variables and generate 1,000 lognormal random variables in Visual Basic for Applications with $\mu = 0$ and $\sigma^2 = 1$, that is, $EY = e^{0.5}$ and $\text{Var}(Y) = e(e - 1)$, go to the Online Supplementary and download the files *Chapter 3 Generate the PDF of Lognormal Random Variables* and *Chapter 3 Generate Lognormal Random Variables*.

It can be seen from Figure 3.1 that a lognormal density can never be negative. Furthermore, it is skewed to the right and has a much thicker tail than a normal random variable.

Before concluding this chapter, we would like to bring the readers' attentions to some existing books written on this subject. In the statistical community, many excellent texts have been written on this subject of simulations, see, for example, Ross (2002) and the references therein. These texts mainly discuss traditional simulation techniques without too much emphasis on finance and risk management. They are more suitable for a traditional audience in statistics.

In finance, there are several closely related texts. A comprehensive treatise on simulations in finance is given in the book by Glasserman (2003). A more succinct treatise on simulations in finance is given by Jaeckel (2002). Both of these books assume a considerable amount of financial background from the readers. They are intended for readers at a more advanced level. A book on simulation based on MAT-LAB is Brandimarte (2006). The survey article by Broadie and Glasserman (1998) offers a succinct account of the essence of simulations in finance. For readers interested in knowing more about the background of risk management, the two special volumes of Alexander (1998), the encyclopedic treatise of Crouchy, Galai, and Mark (2000), and the special volume of Dempster (2002) are excellent sources. The recent monograph of McNeil, Frey, and Embrechts (2005) offers an up-to-date account on topics of quantitative risk management.

The present text can be considered as a synergy between Ross (2002) and Glasserman (2003), but at an intermediate level. We hope that readers with some (but not

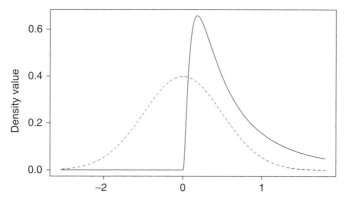

Figure 3.1 Densities of a lognormal distribution with mean $e^{0.5}$ and variance $e(e - 1)$, that is, $\mu = 0$ and $\sigma^2 = 1$ and a standard normal distribution.

highly technical) background in either statistics or finance can benefit from reading this book.

3.5 EXERCISES

1. Verify Equation 3.3.

2. Explain the possible difficulties in implementing quadrature methods to evaluate high dimensional numerical integrations.

3. Using either SPLUS or Visual Basic, simulate 1,000 observations from a lognormal distribution with a mean e^2 and variance $e^4(e^2 - 1)$. Calculate the sample mean and sample variance for these observations and compare their values with the theoretical values.

4. Let a stock have price S at time 0. At time 1, the stock price may rise to S_u with probability p or fall to S_d with probability $(1 - p)$. Let $R_S = (S_1 - S)/S$ denote the return of the stock at the end of period 1.

   (a) Calculate $m_S = E(R_S)$.

   (b) Calculate $v_S = \sqrt{\text{Var}(R_S)}$.

   (c) Let C be the price of a European call option of the stock at time 0 and C_1 be the price of this option at time 1. Suppose that $C_1 = C_u$ when the stock price rises to S_u and $C_1 = C_d$ when the stock price falls to S_d. Correspondingly, define the return of the call option at the end of period 1 as $R_C = (C_1 - C)/C$. Calculate $m_C = E(R_C)$.

   (d) Show that $v_C = \sqrt{\text{Var}(R_C)} = \sqrt{p(1 - p)}(C_u - C_d)/C$.

   (e) Let $\Omega = \frac{(C_u - C_d)}{C} / \frac{(S_u - S_d)}{S}$, the so-called elasticity of the option. Show that $v_C = \Omega v_S$.

The solutions and/or additional exercises are available online at http://www.sta.cuhk .edu.hk/Book/SRMS/.

4

BROWNIAN MOTIONS AND ITÔ'S RULE

4.1 INTRODUCTION

In this chapter, we learn about the notion of Brownian motion and geometric Brownian motion (GBM), the latter being one of the most popular models in financial theory. In addition, the issue of Itô's calculus is also introduced. The key element of this last concept is to develop an operational understanding of Itô's calculus so that readers will be able to do simple stochastic integration such as $\int_0^1 W^2(t)\, dW(t)$. Finally, we learn how to simulate these processes and study their corresponding features.

4.2 WIENER AND ITÔ'S PROCESSES

Consider the model defined by

$$W(t_{k+1}) = W(t_k) + \epsilon_{t_k} \sqrt{\Delta t}, \qquad (4.1)$$

where $t_{k+1} - t_k = \Delta t$, and $k = 0, \dots, N$ with $t_0 = 0$. In this equation, $\epsilon_{t_k} \sim N(0, 1)$ are identical and independent distributed (i.i.d.) random variables. Furthermore, assume that $W(t_0) = 0$. This is known as the random walk model (except for the factor $\sqrt{\Delta t}$, this equation matches with the familiar random walk model introduced in elementary

Simulation Techniques in Financial Risk Management, Second Edition. Ngai Hang Chan and Hoi Ying Wong.
© 2015 John Wiley & Sons, Inc. Published 2015 by John Wiley & Sons, Inc.

courses). Note that from this model, for $j < k$,

$$W(t_k) - W(t_j) = \sum_{i=j}^{k-1} \epsilon_{t_i} \sqrt{\Delta t}.$$

There are a number of consequences as follows:

1. As the right-hand side is a sum of normal random variables, it means that $W(t_k) - W(t_j)$ is also normally distributed.
2. By taking expectations, we have

$$E(W(t_k) - W(t_j)) = 0,$$

$$\text{Var}(W(t_k) - W(t_j)) = E[\sum_{i=j}^{k-1} \epsilon_{t_i} \sqrt{\Delta t}]^2 = (k-j)\Delta t = t_k - t_j.$$

3. For $t_1 < t_2 \le t_3 < t_4$,

$$W(t_4) - W(t_3) \text{ is uncorrelated with } W(t_2) - W(t_1).$$

Equation 4.1 provides a way to simulate a standard Brownian motion (Wiener process). To see how, consider partitioning $[0, 1]$ into n subintervals each with length $\frac{1}{n}$. For each number t in $[0, 1]$, let $[nt]$ denote the greatest integer part of it. For example, if $n = 10$ and $t = \frac{1}{3}$, then $[nt] = [\frac{10}{3}] = 3$. Now define a stochastic process in $[0, 1]$ as follows. For each t in $[0, 1]$, define

$$S_{[nt]} = \frac{1}{\sqrt{n}} \sum_{i=1}^{[nt]} \epsilon_i, \tag{4.2}$$

where ϵ_i are i.i.d. standard normal random variables. Clearly,

$$S_{[nt]} = S_{[nt]-1} + \epsilon_{[nt]} \frac{1}{\sqrt{n}}, \tag{4.3}$$

which is a special form of Equation 4.1 with $\Delta t = \frac{1}{n}$ and $W(t) = S_{[nt]}$. Furthermore, we know that at $t = 1$,

$$S_{[nt]} = S_n = \frac{1}{\sqrt{n}} \sum_{i=1}^{n} \epsilon_i,$$

has a standard normal distribution. Also, by the Central Limit Theorem, we know that S_n tends to a standard normal random variable in distribution even if the ϵ_i are only i.i.d. but not necessarily normally distributed. The idea is that by taking the limit as n tends to ∞, the process $S_{[nt]}$ would tend to a Wiener process in distribution.

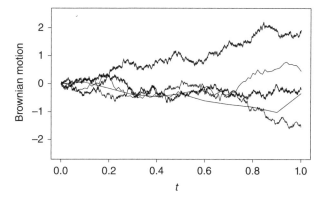

Figure 4.1   Sample paths of the process $S_{[nt]}$ for different n and the same sequence of ϵ_i.

Consequently, to simulate a sample path of a Wiener process, all we need to do is to iterate Equation 4.3. Figure 4.1 shows the simulations on the basis of Equation 4.3.

To generate Figure 4.1 in Visual Basic for Applications, go to the Online Supplementary and download the file *Chapter 4 Generate Brownian Motion Paths with different n.*

To generate Figure 4.2 in Visual Basic for Applications, go to the Online Supplementary and download the file *Chapter 4 Sample paths of Brownian Motion on [0,1].*

In other words, by taking limit as Δt tends to zero, we get a Wiener process (Brownian motion), that is,

$$dW(t) = \epsilon(t)\sqrt{dt},$$

where $\epsilon(t)$ are uncorrelated standard normal random variables. We can interpret this equation as a continuous-time approximation of the random walk model (Eq. 4.1); see Chan (2010). Of course, such an approximation can be dubious because we do not know if this limiting operation is well defined. In more advanced courses in probability, see Billingsley (1999), for example, it is shown that this limiting operation is well defined, and, indeed, we obtain a Wiener process as a limit of the aforementioned operation. Formally, we define a Wiener process $W(t)$ as a stochastic process as follows.

Definition 4.1   *A Wiener process $W(t)$ is a stochastic process that satisfies the following properties:*

- *For $s < t$, $W(t) - W(s)$ is a normally distributed random variable with mean 0 and variance $t - s$.*
- *For $0 \leq t_1 < t_2 \leq t_3 < t_4$, $W(t_4) - W(t_3)$ is uncorrelated with $W(t_2) - W(t_1)$. This is known as the independent increment property.*
- *$W(t_0) = 0$ with probability one.*

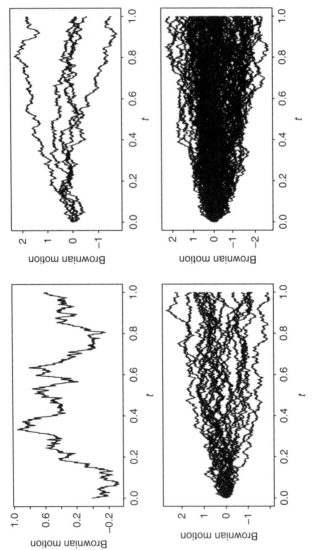

Figure 4.2 Sample paths of Brownian motions on $[0,1]$.

From this definition, we can deduce a number of properties.

1. For $t < s$, $\mathrm{E}(W(s)|W(t)) = \mathrm{E}(W(s) - W(t) + W(t)|W(t)) = W(t)$. This is known as the *martingale* property of the Brownian motion.

2. The process $W(t)$ is nowhere differentiable. Consider

$$\mathrm{E}\left(\left(\frac{W(s) - W(t)}{s - t} \right)^2 \right) = \frac{1}{s - t}.$$

This term tends to ∞ as $s - t$ tends to 0. Hence, the process cannot be differentiable, and we cannot give a precise mathematical meaning to the process $dW(t)/dt$.

3. If we formally represent $\xi(t) = \frac{dW(t)}{dt}$ and call it the white noise process, we can use it only as a symbol, and its mathematical meaning has to be interpreted in terms of an integration in the context of a stochastic differential equation.

The idea of Wiener process can be generalized as follows. Consider a process $X(t)$ satisfying the following equation:

$$dX(t) = \mu\, dt + \sigma\, dW(t), \tag{4.4}$$

where μ and σ are constants, and $W(t)$ is a Wiener process defined previously. If we integrate Equation 4.4 over $[0, t]$, we get

$$X(t) = X(0) + \mu t + \sigma W(t),$$

that is, the process $X(t)$ satisfies the integral equation

$$\int dX(t) = \mu \int dt + \sigma \int dW(t).$$

The process $X(t)$ is also known as a diffusion process or a generalized Wiener process. In this case, the solution $X(t)$ can be written down analytically in terms of the parameters μ and σ and the Wiener process $W(t)$. To extend this idea further, we can let the parameters μ and σ depend on the process $X(t)$ as well. In that case, we have what is known as a general diffusion process or an Itô's process.

Definition 4.2 *An Itô's process is a stochastic process that is the solution to the following stochastic differential equation (SDE):*

$$dX(t) = \mu(x, t)\, dt + \sigma(x, t)\, dW(t). \tag{4.5}$$

In this equation, $\mu(x, t)$ is known as the drift function, and $\sigma(x, t)$ is known as the volatility function of the underlying process. Of course, we need conditions for $\mu(x, t)$ and $\sigma(x, t)$ to ensure Equation 4.5 has a solution. We do not discuss these technical details in this chapter; further details can be found in Karatzas and Shreve (1997) or Dana and Jeanblanc (2002). We will just assume that the drift and the volatility are "nice" enough functions so that the existence of a stochastic process $\{X(t)\}$ that satisfies Equation 4.5 is guaranteed. Again, this equation has to be interpreted through integration.

4.3 STOCK PRICE

Recall the multiplicative model

$$\log S(k+1) = \log S(k) + w(k).$$

The continuous-time version of this equation is

$$d \log S(t) = v \, dt + \sigma \, dW(t).$$

The right-hand side of this equation is normally distributed with mean $v \, dt$ and variance $\sigma^2 dt$. Solving this equation by integration,

$$\log S(t) = \log S(0) + vt + \sigma W(t).$$

Then, $\mathrm{E} \log S(t) = \log S(0) + vt$. As the expected log price grows linearly with t, just as in a continuous compound interest formula, the process $S(t)$ is known as a GBM. Formally, we define

Definition 4.3   *Let $X(t)$ be a Brownian motion with drift v and variance σ^2, that is,*

$$dX(t) = v \, dt + \sigma \, dW(t).$$

The process $S(t) = e^{X(t)}$ is called a GBM with drift parameter μ, where $\mu = v + \frac{1}{2}\sigma^2$. In particular, $S(t)$ satisfies

$$dS(t) = \mu S(t) \, dt + \sigma S(t) \, dW(t),$$

and

$$d \log S(t) = \left(\mu - \frac{1}{2}\sigma^2 \right) dt + \sigma \, dW(t). \tag{4.6}$$

To simulate 1,000 GBMs in Visual Basic for Applications with $\mu = 0.03$ and $\sigma^2 = 0.04$, go to the Online Supplementary and download the file *Chapter 4 Sample path of Geometric Brownian Motion on [0,1]*. A sample path is plotted in Figure 4.3.

Equivalently, $S(t)$ is a GBM starting at $S(0) = z$ if

$$S(t) = ze^{X(t)} = ze^{vt + \sigma W(t)} = ze^{(\mu - \frac{1}{2}\sigma^2)t + \sigma W(t)}.$$

Using this definition, we see that for $t_0 < t_1 < \cdots < t_n$, the successive ratios

$$\frac{S(t_1)}{S(t_0)}, \frac{S(t_2)}{S(t_1)}, \ldots, \frac{S(t_n)}{S(t_{n-1})}$$

are independent random variables by virtue of the independent increment property of the Wiener process. The mean and variance of a geometric Brownian motion can be computed as in the lognormal distribution. Notice that because a Brownian motion is normally distributed, we conclude the following:

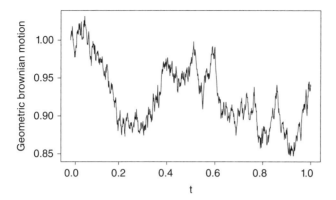

Figure 4.3   Geometric Brownian motion.

1. $\log S(t) = X(t) \sim N(\log S(0) + vt, \sigma^2 t)$.
2. As $S(t) = S(0)e^{X(t)}$,

$$
\begin{aligned}
E(S(t)) = E(E(S(t)|S(0) = z)) &= E(E(ze^{(vt+\sigma W(t))}|S(0) = z)) \\
&= ze^{(\mu-\frac{1}{2}\sigma^2)t}E(e^{\sigma W(t)}) \\
&= ze^{(\mu-\frac{1}{2}\sigma^2)t}E(e^{\sigma\sqrt{t}\xi}) \quad (\xi = W(t)/\sqrt{t} \sim N(0,1)) \\
&= ze^{(\mu-\frac{1}{2}\sigma^2)t}e^{\frac{1}{2}\sigma^2 t} \\
&= ze^{\mu t} = S(0)e^{\mu t}.
\end{aligned}
$$

This equation has an interesting economic implication in the case where μ is positive but small relative to σ^2. On one hand, if $\mu > 0$, then the mean value $E(S(t))$ tends to ∞ as t tends to ∞. On the other hand, if $0 < \mu < \frac{1}{2}\sigma^2$, then the process $X(t) = X(0) + (\mu - \frac{1}{2}\sigma^2)t + \sigma W(t)$ has a negative drift, that is, it is drifting in the negative direction as t tends to ∞. It is intuitively clear that (which can be shown mathematically) $X(t)$ tends to $-\infty$. As a consequence, the original price $S(t) = S(0)e^{X(t)}$ tends to 0. The GBM $S(t)$ is drifting closer to zero as time goes on, yet its mean value $E(S(t))$ is continuously increasing. This example demonstrates the fact that the mean function sometimes can be misleading in describing the process.

3. Similarly, we can show that

$$
Var(S(t)) = S(0)^2 e^{2vt+\sigma^2 t}(e^{\sigma^2 t} - 1) = S(0)^2 e^{2\mu t}(e^{\sigma^2 t} - 1).
$$

4.4   ITÔ'S FORMULA

In the preceding section, we define $S(t)$ in terms of $\log S(t)$ as a Brownian motion. Although such a definition facilitates many of the calculations, it may sometimes be desirable to examine the behavior of the original price process $S(t)$ directly. To see how this can be done, first recall from calculus that

$$d \log S(t) = \frac{dS(t)}{S(t)}.$$

We might be tempted to substitute this elementary fact into Equation 4.6 to get

$$\frac{dS(t)}{S(t)} = v \, dt + \sigma \, dW(t).$$

However, this computation is **NOT** exactly correct because it involves the differential $dW(t)$. A rule of thumb is that whenever we need to substitute quantities regarding $dW(t)$, there is a correction term that needs to be accounted for. We shall provide an argument of this correction term later. For the time being, the correct expression of the previous equation should be

$$\frac{dS(t)}{S(t)} = (v + \frac{1}{2}\sigma^2) \, dt + \sigma \, dW(t)$$

$$= \mu \, dt + \sigma \, dW(t), \qquad (4.7)$$

as $v = \mu - \frac{1}{2}\sigma^2$. The correction term required when transforming $\log S(t)$ to $S(t)$ is known as the **Itô's lemma**. We discuss this in the next theorem. Before doing that, there are a number of remarks.

Remarks

1. The term $dS(t)/S(t)$ can be thought of as the differential return of a stock, and Equation 4.7 says that the differential return possesses a simple form $\mu \, dt + \sigma \, dW(t)$.

2. Note that in Equation 4.7, it is an equation about the ratio $dS(t)/S(t)$. This term can also be thought of as the instantaneous return of the stock. Hence Equation 4.7 is describing the dynamics of the instantaneous return process.

3. In the case of a deterministic dynamics, that is, without the stochastic component $dW(t)$ in Equation 4.7, this equation reduces to the familiar form of a compound return. For example, let $P(t)$ denote the price of a bond that pays $1 at time $t = T$. Assume that the interest rate r is constant over time and there are no other payments before maturity; the price of the bond satisfies

$$\frac{dP(t)}{P(t)} = r \, dt.$$

In other words, $P(t) = P(0)e^{rt} = e^{r(t-T)}$, after taking the boundary condition $P(T) = 1$ into account.

4. Note that Equation 4.7 provides a way to simulate the price process $S(t)$. Suppose we start at t_0 and let $t_k = t_0 + k\Delta t$. According to Equation 4.7, the simulation equation is

$$S(t_{k+1}) - S(t_k) = \mu S(t_k)\Delta t + \sigma S(t_k)\epsilon(t_k)\sqrt{\Delta t},$$

where $\epsilon(t_k)$ are i.i.d. standard normal random variables. Iterating this equation we get

$$S(t_{k+1}) = [1 + \mu\Delta t + \sigma\epsilon_k\sqrt{\Delta t}]S(t_k), \tag{4.8}$$

which is a multiplicative model, but the coefficient is normal rather than lognormal. So this equation does not generate the lognormal price distribution. However, when Δt is sufficiently small, the differences may be negligible.

5. Instead of using Equation 4.7, we can use Equation 4.6 for the log prices and get

$$\log S(t_{k+1}) - \log S(t_k) = \nu\Delta t + \sigma\epsilon(t_k)\sqrt{\Delta t}.$$

This equation leads to

$$S(t_{k+1}) = e^{\nu\Delta t + \sigma\epsilon(t_k)\sqrt{\Delta t}}S(t_k), \tag{4.9}$$

which is also a multiplicative model, but now the random coefficient is lognormal. In general, we can use either Equation 4.8 or Equation 4.9 to simulate stock prices.

With these backgrounds, we are now ready to state the celebrated Itô's lemma, which accounts for the correction term.

Theorem 4.1 *Suppose the random process $x(t)$ satisfies the diffusion equation*

$$dx(t) = a(x, t)\,dt + b(x, t)\,dW(t),$$

where $W(t)$ is a standard Brownian motion. Let the process $y(t) = F(x, t)$ for some function F. Then the process $y(t)$ satisfies the Itô's equation

$$dy(t) = \left(\frac{\partial F}{\partial x}a + \frac{\partial F}{\partial t} + \frac{1}{2}\frac{\partial^2 F}{\partial x^2}b^2\right)dt + \frac{\partial F}{\partial x}b\,dW(t). \tag{4.10}$$

Proof. Observe that if the process is deterministic, ordinary calculus shows that for a function of two variables such as $y(t) = F(x, t)$, the total differential dy is given by

$$dy = \frac{\partial F}{\partial x}dx + \frac{\partial F}{\partial t}dt$$
$$= \frac{\partial F}{\partial x}(a\,dt + b\,dW) + \frac{\partial F}{\partial t}dt.$$

Comparing this expression with Equation 4.10, we see that there is an extra correction term $\frac{1}{2}\frac{\partial^2 F}{\partial x^2}b^2$ in front of dt. To see how this term arises, consider expanding the function F in a Taylor's expansion up to terms of first order in Δt. Note that as ΔW and hence Δx are of order $\sqrt{\Delta t}$, such an expansion would lead to terms with the second order in Δx. In this case,

$$y + \Delta y = F(x, t) + \frac{\partial F}{\partial x}\Delta x + \frac{\partial F}{\partial t}\Delta t + \frac{1}{2}\frac{\partial^2 F}{\partial x^2}(\Delta x)^2$$

$$= F(x, t) + \frac{\partial F}{\partial x}(a\Delta t + b\Delta W) + \frac{\partial F}{\partial t}\Delta t + \frac{1}{2}\frac{\partial^2 F}{\partial x^2}(a\Delta t + b\Delta W)^2.$$

Now focus at the quadratic expression of the last term. When expanded, it becomes

$$a^2(\Delta t)^2 + 2ab(\Delta t)(\Delta W) + b^2(\Delta W)^2.$$

The first two terms of the aforementioned expression are of orders higher than Δt, so they can be dropped as we only want terms up to the order of Δt. The last term $b^2(\Delta W)^2$ is all that remains. Recalling that $\Delta W \sim N(0, \Delta t)$ (recall the earlier fact that $dW(t) = \epsilon(t)\sqrt{dt}$), it can be shown that $(\Delta W)^2 \to \Delta t$. In other words, we have the following approximation

$$dW(t)^2 \cong dt \quad \text{or} \quad dW(t) \cong \sqrt{dt}.$$

Substituting this into the expansion, we have

$$y + \Delta y = F(x, t) + \left(\frac{\partial F}{\partial x}a + \frac{\partial F}{\partial t} + \frac{1}{2}\frac{\partial^2 F}{\partial x^2}b^2\right)\Delta t + \frac{\partial F}{\partial x}b\Delta W.$$

Taking limit as $\Delta t \to 0$ and noting $y(t) = F(x, t)$ complete the proof.   □

Example 4.1   *Suppose $S(t)$ satisfies the geometric Brownian motion equation*

$$dS(t) = \mu S(t)\, dt + \sigma S(t)\, dW(t).$$

Now use Itô's formula to find the equation governing the process $F(S(t)) = \log S(t)$. Using Equation 4.10, we identify $a = \mu S$ and $b = \sigma S$. Furthermore, we know that $\partial F/\partial S = 1/S$ and $\partial^2 F/\partial S^2 = -1/S^2$. According to Equation 4.10, we get

$$d\log S = \left(\frac{a}{S} - \frac{1}{2}\frac{b^2}{S^2}\right)dt + \frac{b}{S}\,dW = (\mu - \frac{1}{2}\sigma^2)\,dt + \sigma\,dW,$$

which agrees with the earlier discussion.

Example 4.2   *Evaluate*

$$\int_0^t s\, dW(s).$$

To evaluate this integral, let us first guess the answer to be the one given by the classical integration by parts formula. That is, we might guess $tW(t) - \int_0^t W(s)\,ds$ to be the answer. To verify it, we need to differentiate this quantity to see if it matches the answer. To do this, use the following steps:

1. Let $X(t) = W(t)$, then $dX(t) = dW(t)$ and we identify $a = 0$ and $b = 1$ in Equation 4.10.
2. Let $Y(t) = F(W(t)) = tW(t)$. Then $\partial F/\partial W = t$, $\partial^2 F/\partial W^2 = 0$, and $\partial F/\partial t = W(t)$.
3. Substitute these expressions into Itô's Lemma, we have $dY(t) = t\,dW(t) + W(t)\,dt$.
4. Integrating the preceding equation, we have

$$Y(t) = \int_0^t s\,dW(s) + \int_0^t W(s)\,ds,$$

that is,

$$\int_0^t s\,dW(s) = tW(t) - \int_0^t W(s)\,ds,$$

as required.

Example 4.3   *Evaluate*

$$\int_0^t W(s)\,dW(s).$$

First guess an answer, $W^2(t)/2$, say. Is this answer correct? To check, we differentiate again and apply Itô's Lemma. Using the recipe,

1. Let $X(t) = W(t)$, then $dX(t) = dW(t)$, and we identify $a = 0$ and $b = 1$ in Equation 4.10.
2. Let $Y(t) = F(W(t)) = W^2(t)/2$. Then $\partial F/\partial W = W$, $\partial^2 F/\partial W^2 = 1$, and $\partial F/\partial t = 0$.
3. Recite Itô's Lemma:

$$dY(t) = \left[\frac{\partial F}{\partial X}a + \frac{\partial F}{\partial t} + \frac{1}{2}\frac{\partial^2 F}{\partial X^2}b^2\right]dt + \frac{\partial F}{\partial X}b\,dW(t),$$

so that

$$dY(t) = \frac{1}{2}dt + W(t)\,dW(t).$$

4. Integrating the preceding equation, we get

$$W^2(t)/2 = Y(t) = \frac{t}{2} + \int_0^t W(s)\,dW(s).$$

In other words,

$$\int_0^t W(s)\,dW(s) = \frac{W^2(t)}{2} - \frac{t}{2}\,!!!$$

5. This time, our initial guess was not correct. We need the extra correction term $\frac{t}{2}$ from Itô's Lemma.

Example 4.4   *Let W_t be a standard Brownian motion and let $Y_t = W_t^3$. Evaluate dY_t.*

Let $X_t = W_t$ and $F(X,t) = X_t^3$. Then the diffusion is $dX_t = dW_t$ with $a = 0$ and $b = 1$. Further

$$\frac{\partial F}{\partial X} = 3X^2, \quad \frac{\partial^2 F}{\partial X^2} = 6X, \quad \frac{\partial F}{\partial t} = 0.$$

Using Itô's lemma, we have

$$dY_t = 3W_t\,dt + 3W_t^2\,dW_t.$$

Integrating both sides of this equation, we get

$$\int_0^t dY_s = \int_0^t 3W_s\,ds + \int_0^t 3W_s^2\,dW_s,$$

$$Y_t = W_t^3 = 3\int_0^t W_s\,ds + 3\int_0^t W_s^2\,dW_s,$$

In other words,

$$\int_0^t W_s^2\,dW_s = \frac{W_t^3}{3} - \int_0^t W_s\,ds.$$

In general, one gets

$$\int_0^t W_s^m\,dW_s = \frac{W_t^{m+1}}{m+1} - \frac{m}{2}\int_0^t W_s^{m-1}\,ds, \quad m = 0, 1, 2, \dots . \qquad (4.11)$$

Example 4.5   *Let*

$$dX_t = \frac{1}{2}X_t\,dt + X_t\,dW_t. \qquad (4.12)$$

Evaluate $d\log X_t$.

From the given diffusion, we have $a = \frac{X_t}{2}$ and $b = X_t$. Let $Y_t = F(X,t) = \log X_t$. Then

$$\frac{\partial F}{\partial X} = \frac{1}{X}, \quad \frac{\partial^2 F}{\partial X^2} = -\frac{1}{X^2}, \quad \frac{\partial F}{\partial t} = 0.$$

Using Itô's lemma, we get $dY_t = d\log X_t = dW_t$. That is, $Y_t = W_t$ Therefore, $X_t = e^{W_t}$ is a solution to Equation 4.12.

Example 4.6 *Let the diffusion be*

$$dX_t = \frac{1}{2} dt + dW_t. \tag{4.13}$$

Evaluate $d\,e^{X_t}$.

From the given diffusion, we have again $a = \frac{1}{2}$ and $b = 1$. Let $Y_t = F(X, t) = e^{X_t}$. Then

$$\frac{\partial F}{\partial X} = e^{X_t}, \quad \frac{\partial^2 F}{\partial X^2} = e^{X_t}, \quad \frac{\partial F}{\partial t} = 0.$$

Using Itô's lemma, we get $dY_t = e^{X_t}\,dt + e^{X_t}\,dW_t$ so that

$$dY_t = Y_t\,dt + Y_t\,dW_t.$$

Example 4.7 *Find the solution to the stochastic differential equation*

$$dX_t = X_t\,dt + dW_t, \quad X_0 = 0.$$

Multiplying the integrating factor e^{-t} to both sides of the SDE, we have

$$e^{-t}\,dX_t = e^{-t}X_t\,dt + e^{-t}\,dW_t.$$

Let $Y_t = e^{-t}X_t$. Then $Y_0 = 0$ and by means of Itô's lemma, we have

$$dY_t = e^{-t}\,dW_t.$$

Integrating both sides of this equation,

$$Y_t - Y_0 = \int_0^t e^{-s}\,dW_s,$$

so that

$$X_t = e^t Y_t = \int_0^t e^{(t-s)}\,dW_s.$$

More generally, if we are given the SDE

$$dX_t = \mu X_t\,dt + \sigma\,dW_t,$$

then using the same method by considering the process $Y_t = e^{-\mu t}X_t$, it can be easily shown that the solution to this SDE is given by the process

$$X_t = \sigma \int_0^t e^{\mu(t-s)}\,dW_s + X_0.$$

Such a process is known as the Ornstein–Uhlenbeck process, which is often used in modeling bond prices.

4.5 EXERCISES

1. Let W_t be a Wiener process. Now is at time t_0. Find the mean and variance of X_t if

(a) $X_t = \sigma_1 \left(W_t - W_{t_2}\right) - \sigma_2 \left(W_{t_1} - W_{t_0}\right)$, $t > t_2 > t_1 > t_0$.

(b) $X_t = \sigma_1 \left(W_t - W_{t_2}\right) - \sigma_2 \left(W_{t_1} - W_{t_0}\right)$, $t > t_1 > t_2 > t_0$.

(c) $X_t = \sum_{j=1}^{n} f(W_{t_{j-1}}) \left(W_{t_j} - W_{t_{j-1}}\right)$, $t_0 < t_1 < \cdots t_n = t$.

(d) Use (c) to show that

$$E\left[\int_0^t f(W_\tau, \tau)\, dW_\tau\right] = 0$$

and

$$E\left[\int_0^t f(W_\tau, \tau)\, dW_\tau\right]^2 = \int_0^t E f(W_\tau, \tau)^2 d\tau.$$

Notice that the aforementioned two identities are known as Itô's identities.

2. Let X_t satisfy the stochastic differential equation

$$dX_t = -\frac{1}{3}dt + \frac{1}{2}dW_t,$$

where $X_0 = 0$ and W_t is a standard Brownian motion process. Define $S_t = e^{X_t}$ so that $S_0 = 1$.

(a) Find the stochastic differential equation that governs S_t.

(b) Simulate 10 independent paths of S_t for $t = 1, \ldots, 30$. Call these paths S_t^i, $i = 1, \ldots, 10$ and plot them on the same graph.

(c) What can you conclude about S_t for t large?

(d) With $n = 10$, evaluate

$$\bar{S}_{30} = \frac{1}{n}\sum_{i=1}^{n} S_{30}^i \tag{4.14}$$

at $t = 30$.

(e) Simulate 100 independent paths and calculate Equation 4.14 with $n = 100$. What can you conclude about $\bar{S}_{1000}$ when n tends to infinity?

3. A stock price is governed by

$$dS(t) = aS(t)\, dt + bS(t)\, dW(t),$$

where a and b are given constants and $W(t)$ is a standard Brownian motion process. Find the stochastic differential equation that governs

$$G(t) = \sqrt{S(t)}.$$

4. Consider a stock price S governed by the geometric Brownian motion process

$$\frac{dS(t)}{S(t)} = 0.10\,dt + 0.30\,dW(t),$$

where $W(t)$ is a standard Brownian motion process.

(a) Using $\Delta t = 1/12$ and $S(0) = 1$, simulate 5,000 years of the process $\log S(t)$ and evaluate

$$\frac{1}{t}\log S(t) \tag{4.15}$$

as a function of t. Note that Equation 4.15 tends to a limit p. What is the theoretical value of p? Does your simulation match with this value?

(b) Evaluate

$$\frac{1}{t}\{\log S(t) - pt\}^2 \tag{4.16}$$

as a function of t. Does this tend to a limit?

5. Simulate a standard Brownian motion process $W(t)$ at grids $0 < \frac{1}{n} < \frac{2}{n} < \cdots < \frac{n-1}{n} < 1$ with $n = 10,000$. Let $W_i = W\left(\frac{i}{n}\right)$ for $i = 0, \ldots, n$ with $W(0) = 0$. Suppose you want to evaluate the integral

$$\int_0^1 W(s)\,dW(s) \tag{4.17}$$

via the approximating sum

$$S_\epsilon = \sum_{i=0}^{n-1}\{(1-\epsilon)W_i + \epsilon W_{i+1}\}\{W_{i+1} - W_i\}. \tag{4.18}$$

(a) On the basis of simulated values of W_i, use Equation 4.18 to evaluate Equation 4.17 with $\epsilon = 0$. Does your result match with the one obtained from Itô's formula?

(b) On the basis of simulated values of W_i, use Equation 4.18 to evaluate Equation 4.17 with $\epsilon = \frac{1}{2}$. This is known as the *Stratonovich integral*. Using your calculated results, can you guess the difference between Itô's integral and the Stratonovich integral?

6. Let W_t denote a standard Brownian motion process.

(a) Let $Y_t = F(W_t) = e^{W_t}$. Write down the diffusion equation that governs Y_t.

(b) Evaluate $\int_0^t e^{W_s}\,dW_s$.

7. Denote X_t as the Brownian motion with drift μ and volatility σ.

(a) Find df and dg where $f(t, X) = tX_t$ and $g(t, X) = tX_t^2$.

(b) Financial market practitioners usually consider the time average of the underlying asset price when making investment decision. If the asset evolves as a Brownian motion X_t, then the time average line can be viewed as a stochastic variable

$$A_t = \frac{1}{t} \int_0^t X_\tau \, d\tau.$$

What is the distribution for A_t?

(c) Suppose $X_0 = 70$, $\mu = 0.5$, and $\sigma = 0.4$. Simulate X_1 and A_1 with $\Delta t = 0.01$. What are the sample means and variances for X_1 and A_1 for 1,000 simulations? What is the covariance between the two random variables, X_1 and A_1?

(d) Comment on your simulation result.

The solutions and/or additional exercises are available online at http://www.sta.cuhk.edu.hk/Book/SRMS/.

5

BLACK–SCHOLES MODEL AND OPTION PRICING

5.1 INTRODUCTION

In this chapter, we apply Itô's Lemma to derive the celebrated option pricing formula by Black and Scholes (1973) in the early 1970s. This formula has far-reaching consequences and plays a fundamental role in modern option pricing theory. Immediately after Black and Scholes, Merton (1973) strengthened and improved the option pricing theory in several ways. To recognize their contributions, Merton and Scholes were awarded the Nobel prize in economics in 1997.

What is an option? An option is a financial derivative (contingent claim) that gives the holder the right (but not the obligation) to buy or to sell an asset for a certain price by a certain date. The option that gives the holder a purchasing right is termed a *call option*, whereas the *put option* gives the holder the selling right. The price in the contract is known as the *exercise price* or *strike price* (K); the date is known as the expiration or maturity (T). American options can be exercised at any time up to expiration. European options can be exercised only on the expiration date. As option holders are given a right, they have to pay an option premium to enter the contract. This premium is usually known as the option price.

Four basic option positions are possible:

1. A long position in a call option. Payoff = $\max(S_T - K, 0)$.
2. A long position in a put option. Payoff = $\max(K - S_T, 0)$.

Simulation Techniques in Financial Risk Management, Second Edition. Ngai Hang Chan and Hoi Ying Wong.
© 2015 John Wiley & Sons, Inc. Published 2015 by John Wiley & Sons, Inc.

3. A short position in a call option. Payoff $= - \max(S_T - K, 0)$.
4. A short position in a put option. Payoff $= - \max(K - S_T, 0)$.

Notice that the long position in a put option is different from the short position of a call option. A long position in an option always has a non-negative payoff, whereas a short position in an option always has a nonpositive payoff, but the option premium is collected up front. Option pricing means determining the correct option premium.

To illustrate the Black–Scholes formula, we first discuss some fundamental concepts in a one period binomial model from which a risk-neutral argument is introduced.

5.2 ONE PERIOD BINOMIAL MODEL

Consider a binomial model in one period. Let S_0 and f denote the initial price of one share of a stock and an option on the stock. After one period, the price of the stock can be either uS_0 or dS_0, where $u > 1$ designates an upward movement of the stock price and $d < 1$ designates a downward movement of the stock price. Correspondingly, the payoff of the option after one period can be either f_u or f_d depending on whether the stock moves up or down. For instance, $f_u = \max(Su - K, 0)$ and $f_d = \max(Sd - K, 0)$ for a call option. Schematically, the one period outcome can be represented by Figure 5.1.

Now consider constructing a hedging portfolio as follows. Suppose that we long (buy and hold) Δ shares of the stock and short (sell) one call option (European). Suppose that the option lasts for one period T and, during the life of the option, the stock can move either up from S_0 to uS_0 or down from S_0 to dS_0. Furthermore, suppose that the risk-free rate in this period is denoted by r. The value of this hedging portfolio in the next period is

$$\Delta uS_0 - f_u, \quad \text{if stock moves up,}$$
$$\Delta dS_0 - f_d, \quad \text{if stock moves down.}$$

This portfolio will be risk free if Δ is chosen so that the value of this portfolio is the same at the end of one period regardless of the stock going up or down, that is,

$$\Delta uS_0 - f_u = \Delta dS_0 - f_d.$$

Figure 5.1 One period binomial tree.

Solving for Δ, we get

$$\Delta = \frac{f_u - f_d}{uS_0 - dS_0}.$$

As this portfolio is risk free in the sense that it attains the same value regardless of the outcome of the stock, it must earn the risk-free rate. Otherwise, one could take advantage of an arbitrage opportunity. For example, if the return of this hedging portfolio is larger than the risk-free rate, one could borrow money from the bank to purchase this portfolio and lock in the fixed return. After one period, the proceeds from the portfolio can be used to repay the loan and the arbitrageur pockets the difference. Consequently, the present value of this portfolio must equal $(\Delta uS_0 - f_u)e^{-rT}$. If we let f denote the value of the option at present, then the present value of the portfolio is $S_0\Delta - f$, and according to the no arbitrage assumption,

$$S_0\Delta - f = (\Delta uS_0 - f_u)e^{-rT}.$$

Consequently,

$$\begin{aligned}
f &= S_0\Delta - (\Delta uS_0 - f_u)e^{-rT} \\
&= S_0\Delta(1 - ue^{-rT}) + f_u e^{-rT} \\
&= \frac{f_u - f_d}{u - d}(1 - ue^{-rT}) + f_u e^{-rT} \\
&= e^{-rT}\left[e^{rT}\frac{f_u - f_d}{u - d}(1 - ue^{-rT}) + f_u\right] \\
&= e^{-rT}\left(e^{rT}\frac{f_u - f_d}{u - d} - u\frac{f_u - f_d}{u - d} + f_u\frac{u - d}{u - d}\right) \\
&= e^{-rT}(f_u\frac{e^{rT} - d}{u - d} + f_d\frac{u - e^{rT}}{u - d}) \\
&= e^{-rT}[pf_u + (1 - p)f_d],
\end{aligned}$$

where $p = \frac{e^{rT} - d}{u - d}$. This identity has a very natural interpretation. If we let the value p, just defined as the probability of the stock, move up in a risk-neutral world, then the aforementioned formula simply states the fact that, in the risk-neutral world,

$$f = e^{-rT}\hat{\mathrm{E}}(f) = e^{-rT}(pf_u + (1 - p)f_d),$$

that is, the expected value of the option in one period discounted by the risk-free rate equals the present value of the option. Note that the expected value in this case is denoted by $\hat{\mathrm{E}}$, which is the expectation taken under the new probability measure p.

For this reason, p is known as the risk-neutral probability. The same reasoning can be used to evaluate the stock itself. Note that

$$\hat{E}(S_1) = puS_0 + (1-p)dS_0$$
$$= pS_0(u-d) + dS_0$$
$$= \frac{e^{rT}-d}{u-d}S_0(u-d) + dS_0$$
$$= e^{rT}S_0.$$

In other words, the stock grows as a risk-free rate under the risk-neutral probability (in the risk-neutral world). Therefore, setting the probability of the stock price moving up to be p is tantamount to assuming that the return of the stock grows as the risk-free rate in a risk-neutral world. In a risk-neutral world, all individuals are indifferent to risk and require no compensation for risk. The expected return of all securities is the risk-free interest rate. It is for this reason that such a computation is usually known as the risk-neutral valuation, and it is equivalent to the no arbitrage assumption in general.

Example 5.1   *Suppose the current price of one share of a stock is $20 and in a period of 3 months, the price will be either $22 or $18. Suppose we sold a European call option with a strike price of $21 in 3 months. Let the annual risk-free rate be 12% and let p denote the probability that the stock moves up in 3 months in the risk-neutral world. Note that the payoff of the option is either $f_u = \$1$ if the stock moves up or $f_d = \$0$ if the stock moves down. How much is the option, f, worth today? To find f, we can use the risk-neutral valuation method. Recall that from the aforementioned discussion,*
$$22p + 18(1-p) = 20e^{0.12/4},$$

so that p = 0.6523. Using the expected payoff of the option, we get

$$\hat{E}(f) = pf_u + (1-p)f_d = p + (1-p)0 = p = 0.6523.$$

Therefore, the value of the option for today is

$$f = e^{-rT}\hat{E}(f) = e^{-0.12/4}(pf_u + (1-p)f_d) = 0.633.$$

Alternatively, we can try to solve the same problem using the arbitrage-free argument.

Example 5.2   *With the same parameters as in the preceding example, consider solving for Δ. Firstly, as we want a risk-free profit for the hedging portfolio, we want to* *purchase Δ shares of the stock and short one European call option expiring in 3 months. After 3 months, the value of the portfolio can be either*

$$22\Delta - 1, \text{ if the stock price moves to \$22,}$$

or

$$18\Delta, \text{ if the stock price moves to \$18.}$$

This portfolio is risk free if Δ is chosen so that the value of the portfolio remains the same for both alternatives, that is,

$$22\Delta - 1 = 18\Delta \quad \text{which means} \quad \Delta = 0.25.$$

The value of the portfolio in 3 months becomes

$$22 \times 0.25 - 1 = 4.5 = 18 \times 0.25.$$

By the no arbitrage consideration, this risk-free profit must earn the risk-free interest rate. In other words, the value of the portfolio today must equal the present value of $4.5, that is, $4.5e^{-0.12/4} = 4.367$. If the value of the option today is denoted by f, then the present value of the portfolio equals

$$20 \times 0.25 - f = 4.5e^{-0.12/4} = 4.367.$$

Solving for f gives
$$f = 0.633,$$

which matches with the answer of the preceding example.

In general, this principle can be applied to a multiperiod binomial tree. We do not go into the analysis of a multiperiod model and refer the readers to Chapter 11 of Hull (2006) for further details. For a comprehensive discussion on the discrete-time approach, see Pliska (1997). Although these two examples are illustrated with a call option, by the same token, the same principle can be used to price a put option; again details can be found in Hull (2006).

5.3   THE BLACK–SCHOLES–MERTON EQUATION

The Black–Scholes option pricing equation has initiated modern theory of finance. Its development has triggered an enormous amount of research and revolutionized the practice of finance. The equation was developed under the assumption that the price fluctuation of the underlying security can be described by a diffusion process studied earlier. The logic behind the equation is conceptually identical to the binomial lattice: at each moment two available securities are combined to construct a portfolio that reproduces the local behavior of a contingent claim. Historically, the Black–Scholes theory predates the binomial lattice.

To begin, let S denote the price of an underlying security (stock) governed by a geometric Brownian motion over a time interval $[0, T]$ by

$$dS = \mu S\, dt + \sigma S\, dW,$$ (5.1)

where W is a standard Brownian motion process. Assume further that there is also a risk-free asset (bond) carrying an interest rate r over the time interval $[0, T]$ such that

$$dB = rB\, dt.$$ (5.2)

Consider a contingent claim that is a derivative (call option) of S. The price of this derivative is a function of S and t, that is, let $f(S, t)$ be the price of the claim at time t when the stock price is S. Our goal is to find an equation that models the behavior of $f(S, t)$. This goal is attained by the celebrated Black–Scholes–Merton equation.

Theorem 5.1   *Using the notation just defined, and assuming that the price and the bond are described by the geometric Brownian motion* (Eq. 5.1) *and the compound interest rate model* (Eq. 5.2), *respectively, the price of the derivative of this security satisfies*

$$\frac{\partial f}{\partial t} + \frac{\partial f}{\partial S} rS + \frac{1}{2}\frac{\partial^2 f}{\partial S^2}\sigma^2 S^2 = rf.$$ (5.3)

Proof. The idea of this proof is the same as the binomial lattice. In deriving the binomial model, we form a portfolio with portions of the stock and the bond so that the portfolio exactly matches the return characteristics of the derivative in a period-by-period manner. In the continuous-time framework, the matching is done at each instant. Specifically, by Itô's lemma, recall that

$$df = \left(\frac{\partial f}{\partial S}\mu S + \frac{\partial f}{\partial t} + \frac{1}{2}\frac{\partial^2 f}{\partial S^2}\sigma^2 S^2\right) dt + \frac{\partial f}{\partial S}\sigma S\, dW.$$ (5.4)

This is also a diffusion process for f with drift $\left(\frac{\partial f}{\partial S}\mu S + \frac{\partial f}{\partial t} + \frac{1}{2}\frac{\partial^2 f}{\partial S^2}\sigma^2 S^2\right)$ and diffusion coefficient $\frac{\partial f}{\partial S}\sigma S$.

Construct a portfolio of S and B that replicates the behavior of the derivative. At each time t, we select an amount x_t of the stock and an amount y_t of the bond, giving a total portfolio value of $G(t) = x_t S(t) + y_t B(t)$. We wish to select x_t and y_t so that $G(t)$ replicates the derivative value $f(S, t)$. The instantaneous gain in value of this portfolio due to changes in security prices is

$$\begin{aligned}
dG &= x_t\, dS + y_t\, dB \\
&= x_t(\mu S\, dt + \sigma S\, dW) + y_t rB\, dt \\
&= (x_t \mu S + y_t rB)\, dt + x_t \sigma S\, dW.
\end{aligned}$$ (5.5)

As we want the portfolio gain of $G(t)$ to behave similarly to the gain of f, we match the coefficients of dt and dW in Equation 5.5 to those of Equation 5.4. Firstly, we match the coefficient of dW in these two equations and we get

$$x_t = \frac{\partial f}{\partial S}.$$

Secondly, as $G(t) = x_t S(t) + y_t B(t)$, we get

$$y_t = \frac{1}{B(t)}(G(t) - x_t S(t)).$$

Thirdly, remember we want $G = f$, therefore,

$$y_t = \frac{1}{B(t)}\left(f(S, t) - \frac{\partial f}{\partial S}S(t)\right).$$

Substituting this expression into Equation 5.5 and matching the coefficient of dt in Equation 5.4, we have

$$\frac{\partial f}{\partial S}\mu S + \frac{1}{B(t)}\left(f(S, t) - \frac{\partial f}{\partial S}S(t)\right)rB(t) = \frac{\partial f}{\partial S}\mu S + \frac{\partial f}{\partial t} + \frac{1}{2}\frac{\partial^2 f}{\partial S^2}\sigma^2 S^2.$$

Consequently,

$$\frac{\partial f}{\partial t} + \frac{\partial f}{\partial S}rS + \frac{1}{2}\frac{\partial^2 f}{\partial S^2}\sigma^2 S^2 = rf.$$

$\square$

Remarks

1. If $f(S, t) = S$, then $\frac{\partial f}{\partial t} = 0$, $\frac{\partial f}{\partial S} = 1$, and $\frac{\partial^2 f}{\partial S^2} = 0$ and Equation 5.3 reduces to $rS = rS$ so that $f(S, t) = S$ is a solution to Equation 5.3.
2. As another simple example, consider a bond where $f(S, t) = e^{rt}$. This is a trivial derivative of S, and it can be easily shown that this f satisfies Equation 5.3.
3. In general, Equation 5.3 provides a way to price a derivative by using the appropriate boundary conditions. Consider a European call option with strike price K and maturity T. Let the price be $C(S, t)$. Clearly, this derivative must satisfy

$$C(0, t) = 0,$$

$$C(S, T) = \max(S - K, 0).$$

For a European put option, the boundary conditions are

$$P(\infty, t) = 0,$$

$$P(S, T) = \max(K - S, 0).$$

Other derivatives may have different boundary conditions. For a knock-out option that will be canceled if the underlying asset breaches a prespecified barrier level (H), in addition to the aforementioned conditions, we have an extra boundary condition

$$f(S = H, t) = 0.$$

4. With these boundary conditions, one can try to solve for the function f from the Black–Scholes equation. One problem is that this is a partial differential equation (PDE), and there is no guarantee that an analytical solution exists. Except in the simple case of a European option, one cannot find an analytic formula for the function f. In practice, either simulation or numerical methods have to be used to find an approximate solution.

5. Alternatively, we can derive Equation 5.3 as follows. Construct a portfolio that consists of shorting one derivative and longing $\frac{\partial f}{\partial S}$ shares of the stock. Let the value of this portfolio be Π and let the value of the derivative be $f(S, t)$. Then

$$\Pi = -f + \frac{\partial f}{\partial S} S. \tag{5.6}$$

The change $\Delta \Pi$ in the value of this portfolio in the time interval Δt is given by

$$\Delta \Pi = -\Delta f + \frac{\partial f}{\partial S} \Delta S. \tag{5.7}$$

Recall that S follows a geometric Brownian motion so that

$$\Delta S = \mu S \, \Delta t + \sigma S \, \Delta W.$$

In addition, from Equation 5.4, the discrete version of df is

$$\Delta f = \left(\frac{\partial f}{\partial S} \mu S + \frac{\partial f}{\partial t} + \frac{1}{2} \frac{\partial^2 f}{\partial S^2} \sigma^2 S^2 \right) \Delta t + \frac{\partial f}{\partial S} \sigma S \, \Delta W.$$

Substituting these two expressions into Equation 5.7, we get

$$\Delta \Pi = \left(-\frac{\partial f}{\partial t} - \frac{1}{2} \frac{\partial^2 f}{\partial S^2} \sigma^2 S^2 \right) \Delta t. \tag{5.8}$$

Note that by holding such a portfolio, the random component ΔW has been eliminated completely. Because this equation does not involve ΔW, this portfolio must equal to the risk-free rate during the time Δt. Consequently,

$$\Delta \Pi = r \Pi \, \Delta t,$$

where r is the risk-free rate. In other words, using Equation 5.8 and Equation 5.6, we obtain

$$\left(\frac{\partial f}{\partial t} + \frac{1}{2}\frac{\partial^2 f}{\partial S^2}\sigma^2 S^2\right)\Delta t = r(f - \frac{\partial f}{\partial S}S)\,\Delta t.$$

Therefore,

$$\frac{\partial f}{\partial t} + \frac{\partial f}{\partial S}rS + \frac{1}{2}\frac{\partial^2 f}{\partial S^2}\sigma^2 S^2 = rf.$$

It should be noted that the portfolio used in deriving Equation 5.3 is not permanently risk free. It is risk free only for an infinitesimally short period of time. As S and t change, $\frac{\partial f}{\partial S}$ also changes. To keep the portfolio risk free, we have to change the relative proportions of the derivative and the stock in the portfolio continuously.

Example 5.3 *Let f denote the price of a forward contract on a non-dividend-paying stock with delivery price K and delivery date T. Its price at time t is given by*

$$f(S, t) = S - Ke^{-r(T-t)}. \tag{5.9}$$

Hence,

$$\frac{\partial f}{\partial t} = -rKe^{-r(T-t)}, \quad \frac{\partial f}{\partial S} = 1, \quad and \quad \frac{\partial^2 f}{\partial S^2} = 0.$$

Substituting these into Equation 5.3, *we get*

$$-rKe^{-r(T-t)} + rS = rf.$$

Thus, the price formula of f given by Equation 5.9 *is a solution of the Black–Scholes equation, indicating that* Equation 5.9 *is the correct formula.*

The Black–Scholes equation generates two important insights. The first one is the concept of risk-neutral pricing. As the Black–Scholes equation does not involve the drift, μ, of the underlying asset price, the option pricing formula should be independent of the drift. Therefore, individual preferences toward the performance or the trend of a particular asset price does not affect the current price of the option on that asset. The second insight is that one would be able to derive a price representation of a European option with any payoff function from the equation. It is summarized in the following theorem.

Theorem 5.2 *Consider a European option with payoff $F(S)$ and expiration time T. Suppose that the continuous compounding interest rate is r. Then, the current European option price is determined by*

$$f(S, 0) = e^{-rT}\widehat{E}[F(S_T)], \tag{5.10}$$

where $\hat{\mathrm{E}}$ denotes the expectation under the risk-neutral probability that is derived from the risk-neutral process

$$\frac{dS}{S} = r\,dt + \sigma\,dW(t). \tag{5.11}$$

Proof. Notice that the current price of the option $f(S,0)$ is a deterministic function of time $t = 0$ and the current asset price S. Consider a stochastic process $\{X_t\}$ that satisfies

$$X_0 = S \quad \text{and} \quad \frac{dX_t}{X_t} = r\,dt + \sigma\,dW(t).$$

Then, $f(S,0) = f(X,0)$. Consider the process $f(X,t)$ derived from the stochastic process of $\{X_t\}$. By Itô's lemma, the differential form of f is

$$df = \left(\frac{\partial f}{\partial t} + rX\frac{\partial f}{\partial X} + \frac{1}{2}\sigma^2 X^2 \frac{\partial^2 f}{\partial X^2} \right) dt + \sigma X \frac{\partial f}{\partial X}\,dW.$$

The Black–Scholes equation says that the coefficient of dt is identical to the term rf; see Theorem 5.1. The total differential for the pricing function is simplified as

$$df = rf\,dt + \sigma X \frac{\partial f}{\partial X}\,dW,$$

which implies

$$df - rf\,dt = \sigma X \frac{\partial f}{\partial X}\,dW.$$

The left-hand side of the aforementioned equation can be combined with the product rule of differentiation to yield

$$e^{rt}\,d\left[e^{-rt} f(X,t) \right] = \sigma X \frac{\partial f}{\partial X}\,dW.$$

This expression has an equivalent integration form,

$$e^{-rT} f(X_T, T) - f(X,0) = \sigma \int_0^T e^{-rt} X \frac{\partial f}{\partial X}\,dW.$$

The right-hand side is a sum of Gaussian processes so that it has an expected value of zero. After taking expectation on both sides,

$$\hat{\mathrm{E}}\left[e^{-rT} f(X_T, T) - f(X,0) \right] = 0.$$

This implies

$$f(X,0) = e^{-rT}\,\hat{\mathrm{E}}[f(X_T, T)].$$

By the terminal condition specified in the Black–Scholes equation, $f(X_T, T) = F(X_T)$, the payoff of the option contract. Hence, we have

$$f(S, 0) = e^{-rT}\,\hat{E}[F(X_T)],$$

where the expectation with respect to the random variable X_T is called the risk-neutral expectation, and the process $\{X_t\}$ is called the risk-neutral asset dynamics. To avoid confusion, financial economists always use the term "asset price process in the risk-neutral world (S_t)" to represent the X_t in this proof. It establishes Equations 5.10 and 5.11 and completes the proof.   □

5.4   BLACK–SCHOLES FORMULA

We are now ready to state the pricing formula of a European call option. A corresponding formula can also be deduced for a European put option. We first establish a key fact about lognormal random variables.

Lemma 5.1 *Let S be a lognormally distributed random variable such that $\log S \sim N(m, v^2)$ and let $K > 0$ be a given constant. Then*

$$E(\max\{S - K, 0\}) = E(S)\Phi(d_1) - K\Phi(d_2), \tag{5.12}$$

where $\Phi(\cdot)$ denotes the distribution function of a standard normal random variable and

$$d_1 = \frac{1}{v}(-\log K + m + v^2) = \frac{1}{v}\left(\log E\left(\frac{S}{K}\right) + \frac{v^2}{2}\right),$$

$$d_2 = \frac{-\log K + m}{v} = \frac{1}{v}\left(\log E\left(\frac{S}{K}\right) - \frac{v^2}{2}\right).$$

Proof. Let $g(s)$ denote the p.d.f. (probability distribution function) of the random variable S. Then

$$E(\max(S - K, 0)) = \int_0^\infty \max(s - K, 0)g(s)\,ds = \int_K^\infty (s - K)g(s)\,ds.$$

By definition, as $\log S \sim N(m, v^2)$,

$$E(S) = e^{(m + \frac{1}{2}v^2)} \quad \text{so that} \quad \log E(S) = m + \frac{1}{2}v^2.$$

Define the variable Q as

$$Q = \frac{\log S - m}{v} \quad \text{so that} \quad Q \sim N(0, 1).$$

The p.d.f. of Q is given by $\phi(q) = \frac{1}{\sqrt{2\pi}}e^{-\frac{q^2}{2}}$, the p.d.f. of a standard normal random variable. Since $q = \frac{\log s - m}{v}$, $s = e^{m+qv}$ so that $dq = \frac{ds}{sv}$. Therefore,

$$
\begin{aligned}
E(\max(S - K, 0)) &= \int_K^\infty \max(s - K, 0)g(s)\,ds \\
&= \int_{\frac{1}{v}(\log K - m)}^\infty (e^{m+qv} - K)g(e^{m+qv})sv\,dq \\
&= \int_{\frac{1}{v}(\log K - m)}^\infty (e^{m+qv} - K)\phi(q)\,dq \\
&= \int_{\frac{1}{v}(\log K - m)}^\infty e^{m+qv}\,\phi(q)\,dq - K\int_{\frac{1}{v}(\log K - m)}^\infty \phi(q)\,dq \\
&= I - II.
\end{aligned}
\tag{5.13}
$$

Note that the third equality follows from the fact that the p.d.f. g of a lognormal random variable S has the form

$$
g(s) = \phi\left(\frac{\log s - m}{v}\right)\frac{1}{sv}, \quad \text{so that } g(e^{m+qv})sv = \phi(q).
\tag{5.14}
$$

We now analyze each of the terms I and II in Equation 5.13. Consider the first term,

$$
\begin{aligned}
I &= e^{m+\frac{v^2}{2}}\int_{\frac{1}{v}(\log K - m)-v}^\infty \phi(q - v)\,d(q - v) \\
&= e^{m+\frac{v^2}{2}}\left(1 - \Phi\left(\frac{\log K - m}{v} - v\right)\right) \\
&= e^{m+\frac{v^2}{2}}\Phi\left(\frac{-\log K + m}{v} + v\right).
\end{aligned}
$$

For the second term, we have

$$
II = K\int_{\frac{1}{v}(\log K - m)}^\infty \phi(q)\,dq = K\Phi\left(\frac{-\log K + m}{v}\right).
$$

Substituting these two expressions into Equation 5.13, we have

$$
E(\max(S - K, 0)) = e^{m+\frac{v^2}{2}}\Phi\left(\frac{-\log K + m}{v} + v\right) - K\Phi\left(\frac{-\log K + m}{v}\right).
$$

Observe that because $\log E(S/K) = -\log K + m + \frac{v^2}{2}$,

$$\frac{-\log K + m}{v} + v = \frac{-\log K + m + v^2}{v}$$

$$= \frac{1}{v}\left(\log E(S/K) + \frac{v^2}{2}\right)$$

$$= d_1.$$

Similarly, it can be easily shown that

$$d_2 = \frac{-\log K + m}{v}.$$

This completes the proof of the lemma. □

Using this lemma, we are now ready to state the Black–Scholes pricing formula.

Theorem 5.3 *Consider a European call option with strike price K and expiration time T. If the underlying stock pays no dividends during the time $[0, T]$ and if there is a continuously compounded risk-free rate r, then the price of this contract at time $0, f(S, 0) = C(S, 0)$, is given by*

$$C(S, 0) = S\,\Phi(d_1) - Ke^{-rT}\Phi(d_2), \tag{5.15}$$

where $\Phi(x)$ denotes the cumulative distribution function of a standard normal random variable evaluated at the point x,

$$d_1 = [\log(S/K) + (r + \sigma^2/2)T]\frac{1}{\sigma\sqrt{T}},$$

$$d_2 = [\log(S/K) + (r - \sigma^2/2)T]\frac{1}{\sigma\sqrt{T}}$$

$$= d_1 - \sigma\sqrt{T}.$$

Proof. The proof of this result relies on the risk-neutral valuation. By Theorem 5.2, we have

$$C(S) = e^{-rT}\hat{E}(\max\{S_T - K, 0\}), \tag{5.16}$$

where S_T denotes the stock price at time T, $\hat{E}$ denotes the risk-neutral expectation, and

$$dS = rS\,dt + \sigma S\,d\hat{W}, \tag{5.17}$$

In this case, we have

$$ES_T = S_0 e^{rT}. \tag{5.18}$$

From the preceding lemma, we get

$$\hat{E}(\max\{S_T - K, 0\}) = \hat{E}(S_T)\Phi(d_1) - K\Phi(d_2).$$

The remaining job is to identify d_1, d_2, and $\hat{E}(S_T)$. By construction, $\hat{E}S_T = S_0 e^{rT}$. Recalling from Equation 5.17, we can easily deduce from Itô's lemma that

$$d \log S_t = \gamma \, dt + \sigma \, d\hat{W}_t, \quad \text{with } \gamma = r - \frac{1}{2}\sigma^2. \tag{5.19}$$

Consequently,

$$m = \hat{E}(\log S_T) = \log S_0 + rT - \frac{1}{2}\sigma^2 T,$$

$$v^2 = \widehat{\text{Var}}(\log S_T) = \sigma^2 T.$$

According to the lemma,

$$d_1 = \frac{-\log K + m + v^2}{v}$$

$$= \frac{1}{\sigma\sqrt{T}}\left[-\log K + \log S_0 + \left(r - \frac{1}{2}\sigma^2\right)T + \sigma^2 T\right]$$

$$= \frac{1}{\sigma\sqrt{T}}\left[\log\left(\frac{S_0}{K}\right) + \left(r + \frac{1}{2}\sigma^2\right)T\right].$$

By similar substitutions, it can be easily shown that

$$d_2 = [\log(S/K) + (r - \sigma^2/2)(T)]\frac{1}{\sigma\sqrt{T}} = d_1 - \sigma\sqrt{T}.$$

This completes the proof of the Black–Scholes formula (Eq. 5.15).   □

Example 5.4   *Consider a 5-month European call option on an underlying stock with a current price of $62, strike price $60, annual risk-free rate 10%, and the volatility of this stock is 20% per year. In this case, $S = 62$, $K = 60$, $r = 0.1$, $\sigma = 0.2$, and $T = \frac{5}{12}$. Applying Equation 5.15, we get*

$$d_1 = \frac{1}{0.2\sqrt{5/12}}\left[\log\left(\frac{62}{60}\right) + \left(0.1 + \frac{0.2^2}{2}\right)\frac{5}{12}\right]$$

$$= 0.641287,$$

$$d_2 = d_1 - 0.2\sqrt{5/12} = 0.512188.$$

From the normal table, we get $\Phi(d_1) = 0.739332$ and $\Phi(d_2) = 0.695740$. Consequently,

$$C = (62)(0.739332) - (60)e^{-(0.1)(5/12)}(0.695740) = 5.798.$$

Remarks

1. Note that the Black–Scholes pricing formula is derived using a risk-neutral valuation argument in this case. Alternatively, for a given derivative such as a European call option, we can try to solve the PDE given by the Black–Scholes Equation 5.3 subject to the explicit boundary conditions given in Remark 3 in Section 5.3. This was the original idea of Black and Scholes, and it is commonly known as the PDE approach. Although feasible, due to the complexity of the PDE of the Black–Scholes equation, the risk-neutral valuation argument offers a more intuitive approach on the basis of the arbitrage-free argument.

2. For a European put option, the corresponding pricing formula is given by

$$P = Ke^{-rT}\Phi(-d_2) - S_0\Phi(-d_1),$$

   where r, K, d_1, and d_2 are defined as in Equation 5.15.

3. To interpret the Black–Scholes formula, look at what happens to d_1 and d_2 as $T \to 0$. If $S_0 > K$, they both tend to ∞ so that $\Phi(d_1) = \Phi(d_2) = 1$ and $\Phi(-d_1) = \Phi(-d_2) = 0$. This means that

$$C = S_0 - K \text{ and } P = 0.$$

   On the other hand, if $S_0 < K$, the reverse argument shows d_1 and d_2 tend to $-\infty$ as $T \to 0$ so that

$$C = 0 \text{ and } P = K - S_0.$$

   Is this reasonable? When $S_0 > K$, and when $T = 0$, the call option should be worth $S_0 - K$ and the put option is of course worthless. On the other hand, if $S_0 < K$ and $T = 0$, the put option should be worth $K - S_0$ and the call option becomes worthless. Thus, the Black–Scholes formula offers the price that is consistent with the boundary condition.

4. What happens when $T \to \infty$? In this case, $d_1 = d_2 = \infty$ and $C = S_0, P = 0$. This is known as the perpetual call. If we own the call for a long time, the stock value will almost certainly increase to a very large value so that the strike price K is irrelevant. Hence, if we own the call we could obtain the stock later for essentially nothing, duplicating the position we would have if we initially bought the stock. Thus, $C = S_0$.

5. The Black–Scholes formula is derived for a European call option under the situation where the stock pays no dividends. When the underlying stock does pay dividends at a specific time during the life of the option, a similar formula to price the option can also be deduced. Again, we refer the interested readers to Hull (2006) for further details.

6. For an American option where early exercise is allowed, one can no longer find an exact analytic formula such as Equation 5.15 for the price of a call. Instead, a range of possible values can be deduced, and details are given in Hull (2006).

7. In using the Black–Scholes formula, one important quantity required is the value of σ, the volatility or the risk of the underlying stock. To use the formula, we can estimate σ from the historical data and put this estimate into the Black–Scholes equation. Such an approach is known as the historical volatility approach. On the other hand, one can also use the Black–Scholes formula to imply the value of σ, known as the implied volatility. In this latter approach, we substitute the observed price of the derivative as the real price into the Black–Scholes formula to solve for σ, giving it the name of implied volatility. This quantity can be used to monitor the market's opinion about the volatility of a particular stock. Analysts often calculate implied volatilities from actively traded options on a certain stock and use them to calculate the price of a less actively traded option on the same stock.

5.5   EXERCISES

1. A company's share price is now $60. Six month from now, it will be either $75 with risk-neutral probability 0.7 or $50 with risk-neutral probability 0.3. A call option exists on the stock that can be exercised only at the end of 6 months with exercise price of $65.

   (a) If you wish to establish a perfectly hedged position, what would you do?

   (b) Under each of the two possibilities, what will be the value of your hedged position?

   (c) What is the expected value of option price at the end of the period?

   (d) What is the reasonable option price today?

2. Consider the binomial model of Section 5.2.

   (a) Show that the European call option price of the two period model is given by

   $$c_2 = \left[p^2 c_{uu} + 2p(1-p)c_{ud} + (1-p)^2 c_{dd}\right] e^{-rT},$$

   where T is the option maturity and

   $$\begin{aligned} c_{uu} &= \max(Su^2 - K, 0) \\ c_{ud} &= \max(Sud - K, 0) \\ c_{dd} &= \max(Sd^2 - K, 0). \end{aligned}$$

   (b) Show by induction that the n-period call price is given by

   $$c_n = e^{-rT} \sum_{j=0}^{n} \left\{ {}_nC_j q^j (1-q)^{n-j} \max\left(Su^j d^{n-j} - K, 0\right) \right\}.$$

   (c) Cox, Ross, and Rubinstein (CRR, 1979) propose that $u = e^{\sigma \sqrt{\Delta t}}$ and $d = e^{-\sigma \sqrt{\Delta t}}$, where σ is the annualized asset volatility, are respectively appropriate choices for the upward and downward factors in implementing the binomial model. Show that

$$\lim_{n\to\infty} c_n = S\,\Phi(d_1) - Ke^{-rT}\Phi(d_2),$$

the Black–Scholes call price, if the CRR proposal is adopted.

3. By Theorem 5.2, show the put-call parity relation

$$p + S = c + Ke^{-rT}.$$

4. A fixed strike geometric Asian call option has the payoff function $\max(G_T - K, 0)$ where

$$G_T = \exp\left(\frac{1}{T}\int_0^T \log S_t\, dt\right).$$

By Theorem 5.2 and Lemma 5.1, determine the analytical solution for the fixed strike geometric Asian call option. (Hints: 1. Apply the result of question 7(b) of Chapter 4, 2. You can find the answer in Chapter 9.)

5. Consider the PDE:

$$\frac{\partial f}{\partial t} + \frac{1}{2}\sigma^2(t,x)\frac{\partial^2 f}{\partial x^2} + \mu(t,x)\frac{\partial f}{\partial x} + a(t,x)f = 0,$$
$$f(T,x) = F(x).$$

By modifying the proof of Theorem 5.2, show that

$$f(t,x) = \mathrm{E}\left[e^{\int_t^T a(\tau)\,d\tau}F(X_T)\right],$$

where X_T is the solution to the SDE:

$$dX = \mu(\tau,X)\,d\tau + \sigma(\tau,X)\,dW_\tau,\quad X_t = x.$$

This result is called the Feynman–Kac formula.

6. Suppose the risk-free interest rate and the volatility of an asset are deterministic functions of time. That means,

$$r = r(t)\quad\text{and}\quad \sigma = \sigma(t).$$

(a) Show that the Black–Scholes equation governing European option prices, $f(t,S)$, is given by

$$\frac{\partial f}{\partial t} + \frac{1}{2}\sigma^2(t)S^2\frac{\partial^2 f}{\partial S^2} + r(t)S\frac{\partial f}{\partial S} - r(t)f = 0.$$

(b) Show that the European call option price satisfies:

$$f(t, S) = e^{-\int_t^T r(\tau)\,d\tau}\widehat{E}[\max(S_T - K, 0)],$$

where
$$dS_\tau = r(\tau)S_\tau\,d\tau + \sigma(\tau)S_\tau\,dW_\tau, \quad \tau > t, \quad \text{and} \quad S_t = S.$$

Hint: Use the result of question 5.

(c) Hence, show that
$$f(t, S) = C_{BS}(t, S; r = \bar{r}, \sigma = \bar{\sigma}),$$

where C_{BS} is the Black–Scholes formula for call option with constant parameters,

$$\bar{r} = \frac{1}{T-t}\int_t^T r(\tau)\,d\tau \quad \text{and} \quad \bar{\sigma} = \sqrt{\frac{1}{T-t}\int_t^T \sigma^2(\tau)\,d\tau}.$$

7. A stochastic process $X(t)$ is said to be a martingale under a probability measure $\mathcal{P}$ if $E^{\mathcal{P}}[X(t)|X(s), s < \tau] = X(\tau)$, with probability one.

(a) Consider the asset price dynamics under the risk-neutral measure:

$$dS = rS\,dt + \sigma S\,dW.$$

Show that $X(t) = S(t)\,e^{-rt}$ is a martingale.

(b) Denote $C(t, S; T)$ as the Black–Scholes formula for a European call option with maturity T. Show that $Ce^{r(T-t)}$ is a martingale.

The solutions and/or additional exercises are available online at http://www.sta.cuhk .edu.hk/Book/SRMS/.

6

GENERATING RANDOM VARIABLES

6.1 INTRODUCTION

The first stage of simulation is the generation of random numbers. Random numbers serve as the building block of simulation. The second stage of simulation is the generation of random variables on the basis of random numbers. This includes generating both discrete and continuous random variables of known distributions. In this chapter, we study techniques for generating random variables.

6.2 RANDOM NUMBERS

Random numbers can be generated in a number of ways. For example, they were generated manually or mechanically by spinning wheels or rolling dice in the old days. Of course, the notion of randomness may be a subjective judgment. Things that look apparently random may not be random according to the strict definition. The modern approach is to use a computer to generate pseudo-random numbers successively. These pseudo-random numbers, although deterministically generated, constitute a sequence of values having the appearance of uniformly $(0, 1)$ distributed random variables.

One of the most popular devices to generate uniform random numbers is the congruential generator. Starting with an initial value x_0, called the seed, the computer

Simulation Techniques in Financial Risk Management, Second Edition. Ngai Hang Chan and Hoi Ying Wong.
© 2015 John Wiley & Sons, Inc. Published 2015 by John Wiley & Sons, Inc.

successively calculates the values $x_n, n \geq 1$ via

$$x_n = ax_{n-1} + c \text{ modulo } m, \tag{6.1}$$

where a, c, and m are given positive integers, and the equality means that the value $ax_{n-1} + c$ is divided by m and the remainder is taken as the value of x_n. Each x_n is either $0, 1, \ldots, m-1$, and the quantity $\frac{x_n}{m}$ is taken as an approximation to the values of a uniform $(0, 1)$ random variable. As each of the numbers x_n assumes one of the values of $0, 1, \ldots, m-1$, it follows that after some finite number of generated values, a value must repeat itself. For example, if we take $a = c = 1$ and $m = 16$, then

$$x_n = x_{n-1} + 1 \text{ modulo } 16.$$

With $x_0 = 1$, then the range of x_n is the set

$$\{0, 1, 2, 3, 4, 5, 6, 7, 8, 9, 10, 11, 12, 13, 14, 15, 0, \ldots\}.$$

When $a = 5$, $c = 1$, and $m = 16$, then the range of x_n becomes

$$\{0, 1, 6, 15, 12, 13, 2, 11, 8, 9, 14, 7, 4, 5, 10, 3, 0, \ldots\}.$$

We usually want to choose a and m such that for any given seed x_0, the number of variables that can be generated before repetition occurs is large. In practice, one may choose $m = 2^{31} - 1$ and $a = 7^5$, where the number 31 corresponds to the bit size of the machine.

Any set of pseudo-random numbers will by definition fail on some problems. It is therefore desirable to have a second generator available for comparison. In this case, it may be useful to compare results for a fundamentally different generator.

From now on, we will assume that we can generate a sequence of random numbers that can be taken as an approximation to the values of a sequence of independent uniform $(0, 1)$ random variables. We do not explore the technical details about the construction of good generators; interested reader may consult L'Ecuyer (1994) for a survey of random number generators.

6.3   DISCRETE RANDOM VARIABLES

A discrete random variable X is specified by its probability mass function given by

$$P(X = x_j) = p_j, \quad j = 0, 1, \ldots, \quad \sum_j p_j = 1. \tag{6.2}$$

To generate X, generate a random number U, which is uniformly distributed in $(0, 1)$ and set

$$X = \begin{cases} x_0 & \text{if} \quad U < p_0, \\ x_1 & \text{if} \quad p_0 \leq U < p_0 + p_1, \\ \vdots \\ x_j & \text{if} \quad \sum_{i=0}^{j-1} p_i \leq U < \sum_{i=0}^{j} p_i, \\ \vdots \end{cases}$$

Recall that for $0 < a < b < 1$, $P(a < U < b) = b - a$. Thus,

$$P(X = x_j) = P\left(\sum_{i=0}^{j-1} p_i \leq U < \sum_{i=0}^{j} p_i \right) = p_j, \tag{6.3}$$

so that X has the desired distribution. Note that if the x_i are ordered so that $x_0 < x_1 < \cdots$ and if F denotes the distribution function of X, then $F(x_k) = \sum_{i=0}^{k} p_i$ and so

$$X \text{ equals to } x_j \text{ if } x_{j-1} \leq F^{-1}(U) < x_j.$$

That is, after generating U, we determine the value of X by finding the interval $[F(x_{j-1}), F(x_j))$ in which U lies. This also means that we want to find the inverse of $F(U)$ and thus the name of inverse transform.

Example 6.1 *Suppose that we want to generate a binomial random variable X with parameters n and p.*

The probability mass function of X is given by

$$p_i = P(X = i) = \frac{n!}{i!(n-i)!} p^i (1-p)^{n-i}, \ i = 0, 1, \ldots, n.$$

From this probability mass function, we see that

$$p_{i+1} = \frac{n-i}{i+1} \frac{p}{1-p} p_i.$$

The algorithm goes as follows:

1. Generate U.
2. If $U < p_0$, set $X = 0$ and stop.
3. If $p_0 < U < p_0 + p_1$, set $X = 1$ and stop.
   $\vdots$
4. If $p_0 + \cdots + p_{n-1} < U < p_0 + \cdots + p_n$, set $X = n$ and stop.

Recursively, by letting i be the current value of X, $pr = p_i = P(X = i)$, and $F = F(i) = P(X \leq i)$, the probability that X is less than or equal to i, the aforementioned algorithm can be succinctly written as:

Step 1: Generate U.
Step 2: $c = p/(1-p)$, $i = 0$, $pr = (1-p)^n$, $F = pr$.
Step 3: If $U < F$, set $X = i$ and stop.
Step 4: $pr = [c(n-i)/(i+1)]pr$, $F = F + pr$, $i = i+1$.
Step 5: Go to Step 3.

To generate binomial random variables X with parameters $n = 10$ and $p = 0.7$ in Visual Basic for Applications (VBA), go to the Online Supplementary and download the file *Chapter 6 Generate Binomial Random Variables Bin(10,7)*.

6.4   ACCEPTANCE-REJECTION METHOD

In the preceding example, we see how the inverse transform can be used to generate a known discrete distribution. For most of the standard distributions, we can simulate their values easily by means of standard built-in routines available in standard packages. However, when we move away from standard distributions, simulating values become more involved. One of the most useful methods is the acceptance-rejection algorithm.

Suppose that we have an efficient method, for example, a computer package, to simulate a random variable Y having probability mass function $\{q_j, j \geq 0\}$. We can use this as a basis for simulating a distribution X having probability mass function $\{p_j, j \geq 0\}$ by first simulating Y and then accepting this simulated value with a probability proportional to p_Y/q_Y. Specifically, let c be a constant such that

$$\frac{p_j}{q_j} \leq c \text{ for all } j \text{ such that } p_j > 0.$$

Then we can simulate the values of X having probability mass function $p_j = P(X = j)$ as follows:

Step 1: Simulate the value of Y from q_j.
Step 2: Generate a uniform random number U.
Step 3: If $U < \frac{p_Y}{c q_Y}$, set $X = Y$ and stop. Otherwise, go to Step 1.

Theorem 6.1   *The acceptance-rejection algorithm generates a random variable X such that*
$$P(X = j) = p_j, \quad j = 0, 1, \dots.$$

In addition, the number of iterations of the algorithm needed to obtain X is a geometric random variable with mean c.

Proof. First consider the probability that a single iteration produces the accepted value j. Note that

$$P(Y = j, \text{ it is accepted}) = P(Y = j)P(\text{accepted}|Y = j)$$
$$= q_j P(U \le p_j/(cq_j))$$
$$= q_j p_j/(cq_j)$$
$$= p_j/c.$$

Summing over j, we get the probability that a generated random variable is accepted as

$$P(\text{accepted}) = \sum_j p_j/c = 1/c.$$

As each iteration independently results in an accepted value with probability $1/c$, the number of iterations needed is geometric with mean c. Finally,

$$P(X = j) = \sum_n P(j \text{ accepted on iteration } n)$$
$$= \sum_n (1 - 1/c)^{n-1} p_j/c = p_j. \qquad \square$$

Example 6.2 *Suppose that we want to simulate a random variable X taking values in $\{1, 2, \ldots, 10\}$ with probabilities as follows:*

i	1	2	3	4	5	6	7	8	9	10
$P(X = i)$	0.11	0.12	0.09	0.08	0.12	0.1	0.09	0.11	0.07	0.11

Using the acceptance-rejection method, first generate discrete uniform random variables over the integers $\{1, \ldots, 10\}$. That is, $P(Y = j) = q_j = 1/10$ for $j = 1, \ldots, 10$. Firstly, compute the number c by setting $c = \max \frac{p_j}{q_j} = 1.2$. Now generate a discrete uniform random variable Y by letting $Y = [10U_1] + 1$, where $U_1 \sim U(0, 1)$. Then generate another $U_2 \sim U(0, 1)$ and compare if $U_2 \le p_Y/(cq_Y)$. If this condition is satisfied, then $X = Y$ is the simulated value. Otherwise, repeat the steps again.
To generate the random variables and see the code in VBA, go to the Online Supplementary and download the file *Chapter 6 Example 6.2 Generate a RV with Support* $\{1, 2, \ldots, 10\}$.

6.5 CONTINUOUS RANDOM VARIABLES

Generating continuous random variables is very similar to generating discrete random variables. It again relies on two main approaches using uniform random numbers: the inverse transform and the acceptance-rejection method.

6.5.1   Inverse Transform

Theorem 6.2   *Let U be a uniform $(0, 1)$ random variable. For any continuous distribution function F, the random variable X defined by $X = F^{-1}(U)$ has distribution F. In this case,*

$$F^{-1}(u) = \inf\{x \cdot F(x) \geq u\}.$$

Proof. Let F_X denote the distribution of $X = F^{-1}(U)$. Then

$$F_X(x) = P(X \leq x)$$
$$= P(F^{-1}(U) \leq x)$$
$$= P(U \leq F(x))$$
$$= F(x). \qquad\qquad\qquad \square$$

Example 6.3   *Let X be an exponential distribution with rate 1. Then its distribution function is given by $F(x) = 1 - e^{-x}$. Let $x = F^{-1}(u)$, then $u = F(x) = 1 - e^{-x}$, so that $x = -\log(1 - u)$. Thus, we can generate X by generating U and setting $X = -\log(1 - U)$. Moreover, because $(1 - U)$ has the same distribution as U, which is uniform $(0, 1)$, we can simply set $X = -\log U$. Finally, it can be seen easily that if $Y \sim \exp(\lambda)$, then $E(Y) = 1/\lambda$ and $Y = X/\lambda$, where $X \sim \exp(1)$. In this case, we can simulate Y by first simulating U and setting $Y = -\frac{1}{\lambda}\log U$.*

The previous example illustrates how to apply the inverse transform method when the inverse of F can be written down easily. The following example demonstrates the case when the inverse of F is not readily available.

Example 6.4   *Let $X \sim \Gamma(n, \lambda)$. Then it has distribution function*

$$F_X(x) = \int_0^x \frac{\lambda e^{-\lambda y}(\lambda y)^{n-1}}{(n-1)!}\, dy.$$

Clearly, finding the inverse of F_X is not feasible. But recall that $X = \sum_{i=1}^n Y_i$, where $Y_i \sim \Gamma(1, \lambda)$ are i.i.d. (identical and independent distributed). Furthermore, each Y_i has distribution function

$$F_Y(y) = \int_0^y \lambda e^{-\lambda s}\, ds,$$

which is the distribution function of an exponential distribution with rate λ. Therefore, we can generate X via

$$X = -\frac{1}{\lambda}\log U_1 - \cdots - \frac{1}{\lambda}\log U_n = -\frac{1}{\lambda}\log(U_1 \cdots U_n).$$

To generate a random variable X that follows a gamma distribution with parameters $n = 5$ and $\lambda = 10$ in VBA, go to the Online Supplementary and download the file *Chapter 6 Generate Gamma Random Variables*.

The message from these two examples is that, although the inverse transform method is simple, we may need to conduct certain simplifications before applying the method.

6.5.2 The Rejection Method

Suppose that we can simulate from a density g easily. We can use this as a basis to simulate from a density $f(x)$ by first generating Y from g and then accepting the generated value with probability proportional to $f(Y)/g(Y)$. Specifically, let c be such that

$$\frac{f(y)}{g(y)} \leq c \text{ for all } y.$$

Then we generate from f via the following algorithm:

Step 1: Generate Y from a density g.

Step 2: Generate a uniform random number U.

Step 3: If $U \leq \frac{f(y)}{cg(y)} := h(y)$ set $X = Y$, else go to Step 1.

This is exactly the same acceptance-rejection method as in the discrete case. Correspondingly, we have the following result whose proof is almost the same as in the discrete case.

Theorem 6.3 *The random variable X generated by the rejection method has density f. Moreover, the number of iterations that this algorithm needs is a geometric random variable with mean c.*

Proof. Let $f(x) = cg(x)h(x)$, where $c \geq 1$ is a constant, $g(x)$ is also a p.d.f. (probability distribution function) and $0 < h(x) \leq 1$. Let Y have p.d.f. g and $U \sim U(0, 1)$. Consider

$$f_Y(x|U \leq h(Y)) = \frac{P(U \leq h(Y)|Y = x)g(x)}{P(U \leq h(Y))}.$$

For the first part in the numerator, we have

$$P(U \leq h(Y)|Y = x) = P(U \leq h(x)) = h(x).$$

For the denominator, consider

$$P(U \leq h(Y)) = \int P(U \leq h(Y|Y = x))g(x)\,dx$$

$$= \int h(x)g(x)\,dx$$

$$= \int \frac{1}{c}f(x)\,dx$$

$$= \frac{1}{c}.$$

Therefore, $f_Y(x|U \leq h(Y)) = h(x)g(x)c = f(x)$.   □

One of the difficulties in using the rejection method is determining the constant c. Our goal is to find the function $cg(x)$ so that $cg(x) \geq f(x)$ and sample easily from the density $g(x)$. This can be achieved using trial-by-error or, in certain circumstances, by simple analysis, as illustrated in the following example.

Example 6.5   *Suppose that we want to simulate from the density*

$$f(x) = 20x(1 - x)^3, \ 0 < x < 1.$$

First note that f is defined only on the interval (0,1). We may try g that can be simulated easily over the same interval, uniform (0, 1), say, that is, $g(x) = 1, 0 < x < 1$. To determine the smallest number c such that $f(x)/g(x) \leq c$ for all $0 < x < 1$, we first find the maximum value of the ratio $f(x)/g(x) = 20x(1 - x)^3$. Using calculus, differentiating, and setting to zero,

$$\frac{d}{dx}\left(\frac{f(x)}{g(x)}\right) = 0,$$

we solve $x = 1/4$ to be the maximum of f/g. Thus,

$$\frac{f(x)}{g(x)} \leq 20(1/4)(3/4)^3 = 135/64 = c.$$

Therefore,

$$\frac{f(x)}{cg(x)} = 20(64)/(135)x(1 - x)^3.$$

The algorithm becomes:

Step 1: Generate random numbers U_1 and U_2.

Step 2: If $U_2 \leq \frac{256}{27}U_1(1 - U_1)^3$, stop and set $X = U_1$. Else go to Step 1.

To simulate from this distribution in VBA, go to the Online Supplementary and download the file *Chapter 6 Example 6.5 Generate Random Variables using Acceptance Rejection Method.*

6.5.3 Multivariate Normal

An important application of simulation is to handle high dimensional problems. High dimensional problems are usually related to multivariate normal distributions (Gaussian distribution). However, most software packages do not provide algorithms for generating multivariate normal random variables. This section studies algorithms for generating multivariate normal random variables.

A random vector X is said to follow a multivariate normal distribution if all of its elements are normal random variables. The distribution of X is then described as

$$X \sim N(m, \Sigma), \tag{6.4}$$

where $m = E[X]$ is the mean vector and $\Sigma = Var[X]$ is the variance-covariance matrix. Consider a vector $X = (X_1, \dots, X_n)^T$ with $X_i \sim N(\mu_i, \sigma_i^2)$. In this case, the mean vector $m = (\mu_1, \dots, \mu_n)^T$ and the $n \times n$ matrix $\Sigma = [Cov(X_i, X_j)]$, $i, j = 1, \dots, n$.

There is a convenient way to generate a normal random vector X when $\Sigma = I$. $\Sigma = I$ indicates that the elements of X are independent random variables. Therefore, we can generate X_i independently and then stack them up to form the vector X. For a normal random vector with dependent components, that is, $\Sigma \neq I$, decomposition methods are useful.

6.5.3.1 Cholesky Decomposition
The first method is the Cholesky decomposition. Consider two correlated standard normal random variables X_1 and X_2 with correlation coefficient ρ, written as,

$$X = \begin{bmatrix} X_1 \\ X_2 \end{bmatrix} \sim N\left(\begin{bmatrix} 0 \\ 0 \end{bmatrix}, \begin{bmatrix} 1 & \rho \\ \rho & 1 \end{bmatrix} \right).$$

Theorem 6.4 *Correlated random variables X_1 and X_2 can be decomposed into two uncorrelated random variables Z_1 and Z_2 through the linear transformation:*

$$Z_1 = X_1$$
$$Z_2 = \frac{X_2 - \rho X_1}{\sqrt{1 - \rho^2}}.$$

In other words,

$$X = \begin{pmatrix} 1 & 0 \\ \rho & \sqrt{1 - \rho^2} \end{pmatrix} Z, \tag{6.5}$$

where

$$Z \sim N\left(\begin{bmatrix} 0 \\ 0 \end{bmatrix}, \begin{bmatrix} 1 & 0 \\ 0 & 1 \end{bmatrix} \right).$$

Proof. As X_1 and X_2 are linear combinations of normal random variables, they are also normally distributed. Furthermore,

$$E(X_1) = E(X_2) = 0,$$
$$Var(X_1) = 1, \quad Var(X_2) = (1 - \rho^2)\,Var(Z_2) + \rho^2\,Var(Z_1) = 1,$$
$$Cov(X_1, X_2) = Cov(Z_1, Z_2\sqrt{1 - \rho^2} + \rho Z_1) = \rho.$$

Thus, X_1 and X_2 have the desired distribution. □

The linear transformation of Equation 6.5 is called the Cholesky decomposition. It enables us to generate (X_1, X_2) by the following procedures.

Step 1: Generate $Z_1, Z_2 \sim N(0, 1)$ i.i.d..
Step 2: Set $X_1 = Z_1$ and $X_2 = Z_2\sqrt{1 - \rho^2} + \rho Z_1$.

In fact, there is a Cholesky decomposition for $N(m, \Sigma)$. As Σ is a semi-positive definite matrix, that is, $v^T \Sigma v > 0$ for all vector v, there exists a lower triangular matrix L such that $\Sigma = LL^T$. The Cholesky decomposition is an algorithm to obtain this lower triangular matrix L.
For $n \times n$ matrices $\Sigma = [a_{ij}]$ and $L = [l_{ij}]$, the Cholesky decomposition algorithm works as follows.

Step 1: Set $l_{11} = \sqrt{a_{11}}$.
Step 2: For $j = 2, \dots, n$ set $l_{j1} = a_{j1}/l_{11}$.
Step 3: For $i = 2, \dots, n - 1$ conduct Step 4 and Step 5.
Step 4: Set $l_{ii} = \left[a_{ii} - \sum_{k=1}^{i-1} l_{ik}^2 \right]^{1/2}$.
Step 5: For $j = i + 1, \dots, n$, set $l_{ji} = \frac{1}{l_{ii}} \left[a_{ji} - \sum_{k=1}^{i-1} l_{jk} l_{ik} \right]$.
Step 6: Set $l_{nn} = \left[a_{nn} - \sum_{k=1}^{n-1} l_{nk}^2 \right]^{1/2}$.

Given the matrix L, a random vector $X \sim N(m, \Sigma)$ is generated by

$$X = m + LZ, \quad Z \sim N(0, I). \tag{6.6}$$

To perform Cholesky decomposition and generate multivariate normal random variables in VBA, go to the Online Supplementary and download the file *Chapter 6 Cholesky Decomposition*.

Theorem 6.5 *The X obtained in Equation 6.6 follows* $N(m, \Sigma)$.

Proof. The random vector X has a Gaussian distribution as it is a linear combination of Gaussian random variables. Therefore, it suffices to check the mean and variance of X. For the mean,

$$E[X] = m + E[LZ] = m.$$

For the variance,

$$\text{Var}[X] = \text{Var}[LZ] = L\left(\text{Var}[Z]\right)L^T = LL^T = \Sigma.$$

$\square$

Example 6.6   *Consider a portfolio of three assets:* $P(t) = S_1(t) + 2S_2(t) + 3S_3(t)$. *The current assets values are* $S_1(0) = 100, S_2(0) = 60,$ *and* $S_3(0) = 30$. *Suppose that the rate of returns of three assets follows a multivariate normal distribution. Specifically, we let*

$$R_i(t) = \frac{S_i(t + \Delta t) - S_i(t)}{S_i(t)} \quad and \quad R(t) = (R_1(t), R_2(t), R_3(t))^T,$$

where

$$R(t) = \begin{bmatrix} 0.1\Delta t + 0.2\sqrt{\Delta t}X_1 \\ -0.03\Delta t + 0.4\sqrt{\Delta t}X_2 \\ 0.2\Delta t + 0.25\sqrt{\Delta t}X_3 \end{bmatrix},$$

$$\begin{bmatrix} X_1 \\ X_2 \\ X_3 \end{bmatrix} \sim N\left(\begin{bmatrix} 0 \\ 0 \\ 0 \end{bmatrix}, \begin{bmatrix} 1 & -0.1 & 0.2 \\ -0.1 & 1 & 0.1 \\ 0.2 & 0.1 & 1 \end{bmatrix} \right).$$

Simulate 10 sample paths of the portfolio with $\Delta t = 1/100$.

To see the simulation of 10 sample paths of the portfolio in VBA, go to the Online Supplementary and download the file *Chapter 6 Example 6.6 Simulating 10 paths of the portfolio*.

Two graphs are produced by the programme. Figure 6.1 plots 10 portfolio sample paths against time. Figure 6.2 plots one sample path for each individual assets and one sample path of the portfolio. Asset and the portfolio can be identified by their initial values.

6.5.3.2   Eigenvalue Decomposition   The second method is the eigenvalue decomposition. Given an $n \times n$ matrix Σ, if a constant value λ and a *nonzero* vector v satisfy:

$$\Sigma v = \lambda v, \tag{6.7}$$

then λ is called an eigenvalue of the matrix Σ and v is the corresponding eigenvector. In principle, there are n eigenvalues for an $n \times n$ matrix. For the variance-covariance

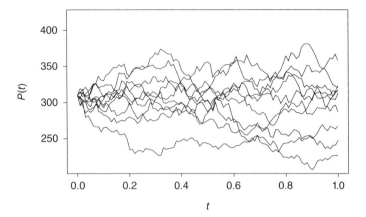

Figure 6.1   Sample paths of the portfolio.

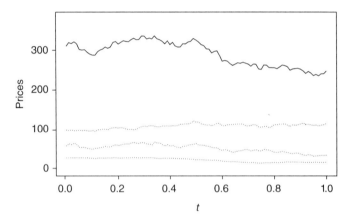

Figure 6.2   Sample paths of the assets and the portfolio.

matrix Σ, we know that all eigenvalues are non-negative and eigenvectors are orthogonal because Σ is semipositive definite.

In multivariate analysis, eigenvalues of a variance-covariance matrix Σ are arranged in descending order as $\lambda_1 > \lambda_2 > \cdots > \lambda_n$ and the corresponding eigenvectors are chosen to have unit length. This means $||v_i|| = 1$ for $i = 1, 2, \ldots, n$. Under these specifications, v_1 is called the first principle component, v_2 is the second principle component, and so on. More importantly, the matrix Σ can be decomposed into a product of three square matrices:

$$\Sigma = PDP^T, \tag{6.8}$$

where $P = [v_1, v_2, \ldots, v_n]$ and $D = diag(\lambda_1, \lambda_2, \ldots, \lambda_n)$ is a diagonal matrix. In SPLUS, eigenvalues and eigenvectors are easily obtained with the subroutine "eigen()".

Theorem 6.6   *If* $Z \sim N(0, I)$, *then* $X = m + P\sqrt{D}Z \sim N(m, \Sigma)$.

Proof.   Again, it suffices to check the mean and variance of X. For the mean,

$$E[X] = m + E[P\sqrt{D}Z] = m.$$

For the variance,

$$\text{Var}[X] = \text{Var}[P\sqrt{D}Z] = P\sqrt{D}\,(\text{Var}[Z])\,[P\sqrt{D}]^T = PDP^T = \Sigma.$$

$\square$

Remarks   VBA users may worry about matrix operations used in the aforementioned algorithms. Fortunately, there are free downloads available on the Web that provide necessary subroutines under the platform of Excel. For instance, the PoPTools, from http://www.cse.csiro.au/poptools/index.htm, includes routines of Cholesky and eigenvalue decompositions.

6.6   EXERCISES

1. Using the inverse transform method to generate a random variable X with the probability mass function.
   (a) $P(X = j) = \frac{1}{j(j+1)}, \quad j = 1, 2, \ldots$.
   (b) $P(X = j) = {}_{(n+j-1)}C_j(1 - p)^j p^n, \quad j = 0, 1, 2\ldots$, where n and p are given parameters.

2. We simulate X, Y, Z from an inverse transform algorithm. Suppose that $U \sim U(0, 1)$. Determine the distributions of the following random variables:
   (a) $X = \text{int}(10U(1 - U))$.
   (b) $Y = \text{int}(1/U)$.
   (c) $Z = (B - 3)^2, \quad B \sim \text{Bin}(5, 0.5)$.

3. Determine the p.d.f. of
   (a) $X = -10 \log U + 5$.
   (b) $X = 2 \tan(\pi U) + 10$.
   (c) $W = nU - \text{int}(nU)$. Show that it is independent of $I = \text{int}(nU)$. (Hint: Show that $P(W \leq w, I = i) = w/n$.)

4. Let X have probability mass function

i	1	2	3	4	5	6	7
$P(X = i)$	0.3	0.12	0.09	0.12	0.1	0.17	0.1

    (a) Use the acceptance-rejection algorithm and simulate 1,000 data points from this distribution. You may use a discrete uniform as your g.

    (b) Plot out the histogram of your simulation.

    (c) What is the expected number of acceptance for this distribution? Does that match your simulation results?

5. Suppose that we want to simulate from the density

$$f(x) = x + 1/2, \ 0 < x < 1.$$

    (a) Using the inverse transformation method, simulate 1,000 values from f.

    (b) Using the acceptance-rejection method, simulate another 1,000 values from f. Which algorithm is more efficient?

6. Suppose that we want to simulate $|Z|$, where $Z \sim N(0, 1)$. That is, the absolute value of a standard normal random variable. First note that the p.d.f. of $|Z|$ is given by

$$f(x) = \frac{2}{\sqrt{2\pi}} e^{-x^2/2}, \ 0 < x < \infty.$$

Suppose you want to use the acceptance-rejection algorithm to simulate $|Z|$. Take g to be the exponential distribution,

$$g(x) = e^{-x}, \ 0 < x < \infty.$$

    (a) Determine the value c such that $c = \max \frac{f(x)}{g(x)}$.

    (b) Use the acceptance-rejection method, simulate 1,000 values of $|Z|$.

    (c) Suppose that you want to recover Z from the simulated values of $|Z|$. One way to do it is to generate a random number U and set

$$Z = \begin{cases} |Z| & \text{if } U > 1/2, \\ -|Z| & \text{if } U \leq 1/2. \end{cases}$$

Using this method, obtain 1,000 values of Z and plot its density.

The solutions and/or additional exercises are available online at http://www.sta.cuhk.edu.hk/Book/SRMS/.

7

STANDARD SIMULATIONS IN RISK MANAGEMENT

7.1 INTRODUCTION

Risk management applications require simulation experiments. In this chapter, we introduce some standard simulation techniques and discuss their applications in risk management.

7.2 SCENARIO ANALYSIS

Scenario analysis of risk management refers to simulating possible scenarios to analyze the risk of a decision and consequences. The ultimate goal of a scenario analysis may be to reach a decision, to verify a model, or to validate a certain conjecture.

Suppose that a newspaper boy buys a newspaper from an agent for \$4 each and sells it for \$6. His problem is to decide how many newspapers to buy each morning. In other words, what would be a prudent purchasing strategy?

To analyze the situation, he examines the sales record for the past 100 days given in Table 7.1. After reviewing the data in Table 7.1, he comes up with the following strategies:

1. Each day, purchase the same number of papers sold the day before.
2. Each day, purchase a fixed number of papers, say 23.

To test each of these two strategies, one could simulate the scenarios using inverse transform. Firstly, convert the information in Table 7.1 into the empirical probability mass function (p.m.f.) (Table 7.2).

Simulation Techniques in Financial Risk Management, Second Edition. Ngai Hang Chan and Hoi Ying Wong.
© 2015 John Wiley & Sons, Inc. Published 2015 by John Wiley & Sons, Inc.

TABLE 7.1   Sales Record

Number of Newspapers	Days Occurring
21	15
22	20
23	30
24	21
25	14

TABLE 7.2   Probability Mass Function

Number of Newspapers	p.m.f.	Cumulative Distribution
21	0.15	0.15
22	0.20	0.35
23	0.30	0.65
24	0.21	0.86
25	0.14	1.00

TABLE 7.3   Policy Simulation and Evaluation

	$u \sim U(0, 1)$	Number of Newspapers	Profit of 1	Profit of 2
Day 1	0.5828	23	$46	$46
Day 2	0.0235	21	$34	$34
Day 3	0.5155	23	$42	$46
Day 4	0.3340	22	$40	$40
Day 5	0.4329	23	$44	$46
Day 6	0.2259	22	$40	$40
Day 7	0.5798	23	$44	$46
Day 8	0.7604	24	$46	$46
Day 9	0.8298	24	$48	$46
Day 10	0.6405	23	$42	$46
		Total profit =	$426	$436

Now simulate 10 future days and compare the two policies following the p.m.f. given in Table 7.2. The simulation draws a standard uniform random variables u. The demands of newspaper are generated according to where the random variables fall. For instance, if $u = 0.17$, which belongs to the range of 0.15–0.35, then the corresponding demand is 22. To have a fair comparison, assume that the newspaper boy orders 23 papers on Day 1. Table 7.3 lists the results of the simulation. The interval [0, 1] is partitioned according to the cumulative frequency in Table 7.2. According to Table 7.3, policy 2 is better than policy 1. One can repeat the simulation for many times to see if this phenomenon is consistent.

The newspaper boy example illustrates several important elements in scenario analysis. Decision makers identify possible scenarios on the basis of empirical data or experience. In this example, scenarios correspond to the daily demand of newspapers. Simulation is then developed to replicate future possibilities. We use the inverse transform with the empirical density function in this example. After generating scenarios, a risk manager analyzes consequences corresponding to each scenario. If the first policy is adopted, then the number of newspapers purchased equals the number sold the previous day; otherwise, 23 papers are purchased. Finally, evaluation and comparison can be conducted using the simulated results.

7.2.1 Value at Risk

In finance, risk scenario analysis is usually conducted for evaluating value-at-risk (VaR), a widely adopted risk measure.

Definition 7.1 *VaR summarizes the worst loss of a portfolio over a target horizon with a given level of confidence.*

Statistically speaking, VaR describes the specified *quantile or percentile* of the projected distribution of profits and losses over the target horizon. Let R_t be the return of a portfolio for a horizon t. Then, the $c\%$ confidence VaR of the portfolio is measured through the expression:

$$P(R_t < -\text{VaR}) = (1 - c)\% := \alpha. \tag{7.1}$$

Hence, VaR is the negative of the α-th percentile of the probability distribution of profits and losses. The larger the VaR, the higher the risk of the portfolio. An advantage of VaR is that it allows the user to specify the confidence level to reflect individual risk-averseness. For more details, see Jorion (2000).

VaR is indispensable for market risk analysis because it is the number that splits future possible asset returns into two scenarios: risky and nonrisky. Returns less than the negative of VaR belong to the class of risky scenario. Decision makers can evaluate their policies by examining consequences under the risky scenario. For instance, a bank may check if it maintains enough money for an extremely risky situation.

A conventional way to measure VaR often assumes portfolio returns to follow a normal distribution. VaR obtained in this way is called normal VaR. A typical model is

$$R_t = \mu + \sigma Z, \quad Z \sim \text{N}(0, 1). \tag{7.2}$$

In such a parametric model, it is easy to derive that

$$\text{VaR}_\alpha(t) = -z_\alpha \sigma - \mu, \tag{7.3}$$

where z_α is the α-quantile of the standard normal distribution, μ is the drift, and σ is the standard deviation of the return R_t over the horizon t.

Although one can prove Equation 7.3 mathematically, we would like to verify it by simulation. The algorithm is given as follows.

Step 1: Generate n independent standard normal random variables, namely $Z_j \sim N(0, 1)$ i.i.d. (identical and independent distributed), $j = 1, 2, \dots, n$.

Step 2: Set $R_j = \mu + \sigma Z_j$.

Step 3: Rank $\{R_1, R_2, \cdots, R_n\}$ in ascending order as $\{R_1^*, R_2^*, \cdots, R_n^*\}$.

Step 4: Set VaR $= -R_k^*$, where $k = \text{int}(\alpha \times n)$.

Example 7.1   *Let $\mu = 0.003$, $\sigma = 0.23$, $\alpha = 5\%$, and $n = 10,000$. Then, the 95% VaR corresponds to the 500th smallest return generated from the simulation. Our simulation shows that the VaR $= 0.3783$, which is close to the value, 0.3753, obtained by Equation 7.3. To see the code in Visual Basic for Applications, go to the Online Supplementary and download the file Chapter 7 Example 7.1.*

7.2.2   Heavy-Tailed Distribution

In reality, returns of market prices may not follow a normal distribution but a heavy-tailed distribution. This means that the two tails of the empirical density decay less rapidly than the normal density. Because closed form solution for the VaR of a heavy-tailed distribution is not readily available, a feasible alternative is to generate random variables according to a heavy-tailed distribution.

One commonly used form for heavy-tailed distribution is the generalized error distribution (GED). The p.d.f. (probability distribution function) of GED with parameter ξ is given by

$$f(z) = \frac{\xi \exp\left(-\frac{1}{2}|z/\lambda|^\xi\right)}{\lambda 2^{1+1/\xi}\Gamma(1/\xi)}, \tag{7.4}$$

$$\lambda = \left[\frac{2^{-2/\xi}\Gamma(1/\xi)}{\Gamma(3/\xi)}\right]^{1/2},$$

where $\Gamma(\cdot)$ denotes the Gamma function. Figure 7.1 plots the p.d.f. of GED, and Figure 7.2 zooms in at the left-tail of the density function. It is seen that the smaller the ξ is, the heavier the left-tail of the density function will be.

The key to simulate VaR is to generate random variables following the desired distribution. In this case, we apply the rejection method introduced in Chapter 6 using an exponential distribution for g. The algorithm goes as follows.

Step 1: Generate $Y \sim \text{Exp}(1)$.

Step 2: Generate $U \sim U(0, 1)$.

Step 3: If $U \le 2f(Y)e^Y/a$, then $Z = Y$; else go to Step 1.

Step 4: Generate $V \sim U(0, 1)$. If $V < 1/2$, then $Z = -Y$.

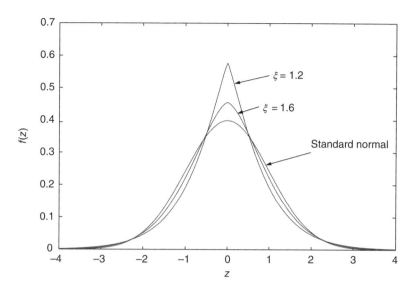

Figure 7.1   The shape of GED density function.

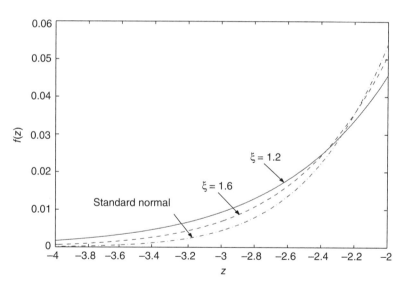

Figure 7.2   Left tail of GED.

Step 5: Repeat Steps 1–4 for n times to get $\{Z_1, Z_2, \ldots, Z_n\}$.

Step 6: Set $R_i = \mu + \sigma Z_i$.

Step 7: Sort the returns in ascending order as $\{R_1^*, R_2^*, \ldots, R_n^*\}$.

Step 8: Set VaR $= -R_k^*$ where $k = \text{int}(\alpha \times n)$.

Remarks

1. In Step 3, a is a constant no less than $\max_y \{2f(y)e^y\}$.
2. As the exponential distribution is defined with a domain of positive real numbers, Steps 1–3 of the algorithm generate positive GED. Step 4 converts a positive GED random variable into a GED random variable.

7.2.3   Case Study: VaR of Dow Jones

We demonstrate the use of GED-VaR by considering 10-year daily closing prices of Dow Jones Industrial Index (DJI) in the period of August 8, 1995, to August 7, 2004. Data downloaded from http://finance.yahoo.com consists of 2,265 prices. The prices are converted into 2,264 daily returns by the formula:

$$R_t = \frac{S_t - S_{t-1}}{S_{t-1}}.$$

Sample mean and standard deviation of the returns are 0.04% and 1.16% in a daily scale, respectively. From Equation 7.3, the 95% and 99% normal VaR from the sample are 1.87% and 2.66%, respectively.

To access the quality of normal VaR, one has to test the normality assumption or, more precisely, the distributional assumption used in the VaR computation. Here, we introduce a simple but valuable tool, known as the quantile–quantile (QQ) plot. The idea is to plot the quantiles of the sample returns against the quantiles of the distribution used. If the returns truly follow the target distribution, then the graph should look similar to a straight line. For testing normality, the target distribution is the normal distribution. Systematic deviations from the line signal that the returns are not well described by the normal distribution.

Figure 7.3 shows a QQ plot of our sample against the normal distribution. Large deviations are observed by the two tails of the empirical data. Specifically, the empirical quantile is less than the normal quantile in the left tail but larger than the normal quantile in the right tail. The deviations strongly suggest heavy-tailed distribution from the empirical data.

We use GED to reduce the deviation from the QQ plot. Returns are first standardized by the sample mean and standard deviation as

$$SR_t = \frac{R_t - 0.04\%}{1.16\%},$$

where SR_t denotes the standardized return at time t. We conjecture that $SR_t \sim GED(\xi)$, identically and independently. The parameter ξ is estimated from the SR using maximum likelihood estimation (MLE). Our estimation shows that $\xi = 1.21$ (Appendix). Then, GED-VaR is estimated from the eight-step algorithm in Section 7.2.1, where the constant a is required in Step 3. The value of a can be deduced from the plot of $2f(y)e^y$ against y, where $f(y)$ is the p.d.f. of GED(1.21).

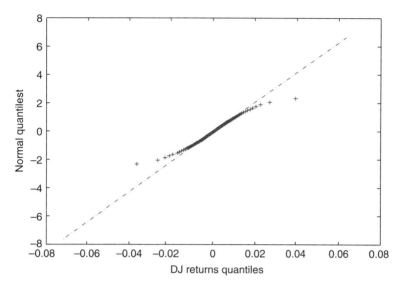

Figure 7.3   QQ plot of normal quantiles against daily Dow Jones returns.

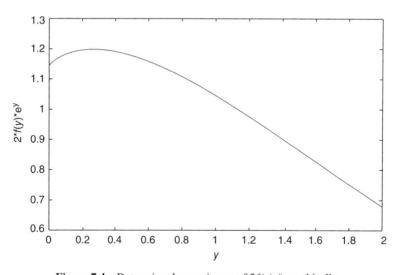

Figure 7.4   Determine the maximum of $2f(y)e^y$ graphically.

Figure 7.4 shows that the maximum function value is bounded above by 1.2 so that we set $a = 1.2$.

To simulate GED-VaR in Visual Basic for Applications, go to the Online Supplementary and download the file *Chapter 7 Simulating GED-VaR*.

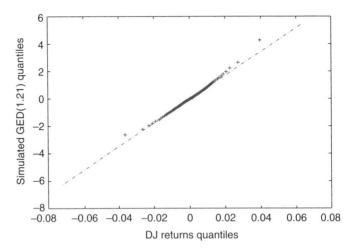

Figure 7.5   QQ plot GED(1.21) quantiles against Dow Jones return quantiles.

The program estimates 95% and 99% VaR by generating 10,000 GED(1.21) random variables. For the confidence intervals, it repeats the process 1,000 times to get 1,000 VaR estimates. After arranging the simulated VaRs in ascending order, the 95% two-tailed confidence interval (CI) is the range between the 25th VaR and the 975th VaR.

To check the performance of GED-VaR, we use the QQ-plot of Figure 7.5 on the basis of one simulation. It is seen that deviations from the straight line have been substantially reduced. From this exercise, we see that GED(1.21) is appropriate for modeling the sample of Dow Jones returns. The average 95% VaR and 99% VaR from the 1,000 simulation are 1.87% and 3.02%, respectively. Therefore, 95% GED-VaR and 95% normal VaR give similar values, whereas 99% GED-VaR is 10% more than the 99% normal VaR.

These findings may be useful for a risk manager. As normal VaR is commonly used in the financial industry, it is essential for a risk manager to understand the limitation of the normal VaR. The rationale of this empirical study is that normal VaR is a good estimate for potential losses of a portfolio under "normal, nonextreme" scenarios. However, it underestimates potential losses when "extreme events" happen, especially for those happening with probability less than 1%. To measure VaR with higher confidence level, for example, 99% VaR, the risk manager may consider GED-VaR. For further discussion about extreme values, see Embrechts, Klüppelberg, and Mikosch (1997) and the themed volume of Finkenstädt and Rootzén (2004).

7.3 STANDARD MONTE CARLO

In the preceding chapters, we studied the idea of simulating random variables. One of the main reasons to simulate random variables is to estimate quantities such as E(X), which is related to the evaluation of definite integrals. Suppose we have already generated n values of a random variable X, it would be very natural to estimate the quantity $\theta = E(X)$ by $\overline{X}_n = \frac{1}{n} \sum_{i=1}^{n} X_i$. We study some standard statistical techniques to assess the accuracy of such an estimate, which are based on the law of large numbers and the central limit theorem. Whenever we estimate quantities such as $E(X)$ on the basis of standard applications of simulations, we refer these methods as standard Monte Carlo simulations. We study other more sophisticated simulation methods in later chapters.

7.3.1 Mean, Variance, and Interval Estimation

Suppose that X is a given random variable with mean θ and variance σ^2. A natural way to evaluate $\theta = E(X)$ using simulations is to generate random values $X_1, \ldots, X_n$ and calculate the quantity

$$\overline{X}_n = \frac{1}{n} \sum_{i=1}^{n} X_i,$$

which is called the sample mean of $\{X_1, \ldots, X_n\}$. It is easy to see that

$$E(\overline{X}_n) = E(X) = \theta, \text{ unbiasedness property,} \tag{7.5}$$

$$Var(\overline{X}_n) = \frac{\sigma^2}{n}. \tag{7.6}$$

To assess the accuracy of $\overline{X}_n$ as an estimate of θ, we rely on two important results. The first one is the law of large numbers, which asserts that as the number of simulations n gets bigger, the closer is $\overline{X}_n$ to θ, see, for example, Casella and Berger (2001). Specifically,

Theorem 7.1 *Let $X_1, \ldots, X_n$ be i.i.d. random variables with mean θ and variance σ^2. Then for any given $\epsilon > 0$,*

$$P(|\overline{X}_n - \theta| > \epsilon) \to 0 \text{ as } n \to \infty.$$

This result is sometimes written as $\overline{X}_n \to \theta$ in probability.

The second one is the central limit theorem, which asserts that as n tends to infinity, the distribution of the random variable $\overline{X}_n$ behaves as a normal distribution approximately.

Theorem 7.2 *Let $X_1, \ldots, X_n$ be i.i.d. random variables with mean θ and variance $\sigma^2 > 0$. Then as n tends to infinity*

$$P(\sqrt{n} \frac{(\overline{X}_n - \theta)}{\sigma} \leq z) \to \Phi(z),$$

where $\Phi(z)$ denotes the c.d.f. (cumulative distribution function) of a standard normal distribution evaluated at the point z.

A equivalent definition of this result is that the random variable $\sqrt{n}(\overline{X}_n - \theta)/\sigma$ converges in distribution to Z, written as

$$\sqrt{n}\frac{(\overline{X}_n - \theta)}{\sigma} \to_{\mathcal{L}} Z,$$

where $Z \sim N(0, 1)$. The proof of these two results can be found in standard text books in probability; see Billingsley (1999) for example. One immediate application of the central limit theorem is to construct approximate confidence intervals for θ. According to Theorem 7.2,

$$P(|\overline{X}_n - \theta| > \frac{\sigma}{\sqrt{n}}c) \sim P(|Z| \geq c) = 2(1 - \Phi(c)).$$

As a result, if we let $c = 1.96$, then the probability of $\overline{X}_n$ differs from θ by more than $1.96\sigma/\sqrt{n}$ would be approximately equal to 0.05. In other words, we are relatively confident (95%) that our estimate is within two standard errors ($1.96\sigma/\sqrt{n}$) from θ. To make use of this result, we have to have knowledge about the value σ, which is usually unavailable. A simple fix is to estimate it from the simulated values. The sample variance, which is defined as

$$S^2 = \frac{1}{n-1} \sum_{i=1}^{n}(X_i - \overline{X}_n)^2,$$

constitutes an estimate of σ^2. It can be easily shown that

$$E(S^2) = \sigma^2, \text{ unbiasedness property,} \tag{7.7}$$

$$S_{j+1}^2 = (1 - 1/j)S_j^2 + (j + 1)(\overline{X}_{j+1} - \overline{X}_j)^2. \tag{7.8}$$

One frequently asked question in simulations is that after simulating X and evaluating $\overline{X}_n$, when should we stop? The answer to this question is given by the following scheme:

1. Choose an appropriate value d for the standard deviation of the estimation. That is, d represents the margin of error we can tolerate using simulations.
2. Generate at least 100 values of X.
3. Continue generating X and stopping when we have k values of X such that $S/\sqrt{k} < d$.
4. The desired estimate is given by $\overline{X}_k$.

Finally, we can form an interval estimation for θ by using the notion of confidence intervals.

Definition 7.2 *If $\overline{X}_n = \overline{x}, S = s$, then the interval*

$$\left(\overline{x} - z_{\alpha/2} \frac{s}{\sqrt{n}}, \overline{x} + z_{\alpha/2} \frac{s}{\sqrt{n}} \right)$$

is an approximate $100(1 - \alpha)\%$ *confidence interval for* θ.

In particular, when $\alpha = 0.05$, $z_{\alpha/2} = 1.96$ and $(\overline{x} \pm 1.96s/\sqrt{n})$ is an approximate 95% confidence interval for θ and thus giving rise to the rule of "two sigma."

7.3.2 Simulating Option Prices

To illustrate the ideas of standard simulations in risk management, consider first simulating stock prices. Let S denote the price of a stock. Recall that we usually assume that S follows a geometric Brownian motion

$$dS = \mu S \, dt + \sigma S \, dW.$$

Equivalently,

$$d \log S = v \, dt + \sigma \, dW,$$

where $v = \mu - \sigma^2/2$. Using the last equation and letting ϵ to denote a standard normal random variable, we can generate S according to the formula

$$S(t + dt) = S(t) \exp(v \, dt + \sigma \epsilon \sqrt{dt}).$$

In particular, $S_T = S_0 e^{X_T}$, where $X_T = vT + \sigma W_T \sim N(vT, \sigma^2 T)$ (Section 4.3), so we have

$$S(T) = S(0) \exp(vT + \sigma \epsilon \sqrt{T}). \tag{7.9}$$

Notice that according to the risk neutral valuation principle, we usually take $\mu = r$, the risk-free rate.

Example 7.2 *Let $S_0 = 10$, $\mu = r = 0.03$, $\sigma = 0.4$, and $dt = 1/52$. We want to simulate weekly prices of the stock $S_i, i = 1, \ldots, 52$ for a 1-year period. Then $v = \mu - \sigma^2/2 = -0.05$ and the results are given in Table 7.4. To see the code in Visual Basic for Applications, go to the Online Supplementary and download the file Chapter 7 Example 7.2 Simulating weekly prices of the stock.*

Suppose that we want to calculate the price of a European call option maturing in 1 year with strike price $K = 12$. We can use the Black–Scholes formula to obtain the call price C as

$$C(S, t) = S\Phi(d_1) - Ke^{-r(T-t)}\Phi(d_2),$$

TABLE 7.4   Simulated Prices of the First and the Last 10 Weeks

Week	Price
0	10.000000
1	10.38419
2	10.37402
3	10.67406
4	11.65342
5	11.89871
6	11.28296
7	11.15327
8	10.33483
9	11.16090
10	12.14546
⋮	⋮
43	14.39009
44	13.78038
45	14.01125
46	12.72393
47	13.44627
48	13.05377
49	12.00424
50	12.74416
51	12.16204
52	12.15517

where $d_1 = \frac{1}{\sigma\sqrt{T-t}}(\log(S/K) + (r + \sigma^2/2)(T - t))$ and $d_2 = d_1 - \sigma\sqrt{T - t}$. Substituting the values of $r = 0.03, K = 12, T = 1, t = 0, \sigma = 0.4$, and $S_0 = 10$, we get

$$d_1 = \frac{1}{0.4\sqrt{1}}(\log(10/12) + (0.03 + 0.08)(1)) = -0.1808, d_2 = d_1 - 0.4 = -0.5808.$$

Using the SPLUS command pnorm(z) to evaluate $\Phi(z)$, we get $\Phi(d_1) = 0.4283$ and $\Phi(d_2) = 0.2807$. Hence,

$$C = 10(0.4283) - 12e^{-0.03}(0.2807) = 1.013918.$$

On the other hand, we can evaluate $C = e^{-rT}\hat{E}(S_T - K)^+$.

Example 7.3   *The price of the European call option can now be computed using simulations.*

1. First generate n independent values of $S_1(T), \ldots, S_n(T)$ according to Equation 7.9.

2. Compute simulated discounted call prices $C_i = e^{-rT} \max\{(S_i(T) - K), 0\}, i = 1, \ldots, n.$

3. Compute $\overline{C} = \frac{1}{n} \sum_{i=1}^{n} C_i$. $\overline{C}$ is an estimate of the discounted payoff $\hat{E}(S_T - K)^+$.

4. Construct a 95% confidence interval for C from

$$\overline{C} \pm 1.96S/\sqrt{n},$$

where

$$S = \sqrt{\frac{1}{n-1} \sum_{i=1}^{n} (C_i - \overline{C})^2},$$

is the sample standard deviation of the simulated call prices C_i.

To simulate 100 paths of stock price to compute call option price and its confidence interval in Visual Basic for Applications, go to the Online Supplementary and download the file *Chapter 7 Example 7.3 Computing European call option price by simulation.*

Outputs of the simulated C_is are given in Table 7.5. The result of a 100-path simulation shows that the 95% confidence interval for C is $(0.37, 1.83)$. Figure 7.6 shows that when the number of runs increases, the value of C converges to the limit of 1.01.

TABLE 7.5   The Discounted Call Prices for the First 20 Paths

Path	C_i
1	0.0000000
2	0.0000000
3	0.0000000
4	5.9331955
5	1.1971242
6	0.0000000
7	2.2395878
8	0.0000000
9	0.0000000
10	4.0065595
11	0.0000000
12	1.3006804
13	0.0000000
14	0.0000000
15	0.0000000
16	0.0000000
17	6.0970236
18	0.0000000
19	0.0000000
20	0.1768191

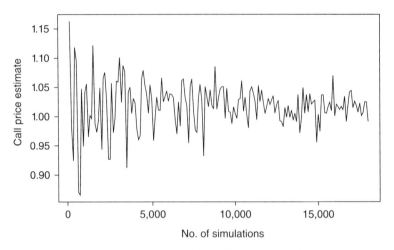

Figure 7.6   Simulations of the call price against the size.

7.3.3   Simulating Option Delta

In risk management, hedging an option is sometimes more important than valuing the option. When a bank issues structured financial products to enhance sales, the embedded option risk would be of great concern. Hedging is a useful device to manage such a risk. For a standard call option, the hedge ratio refers to the delta of the option, the partial derivative of the option price with respect to the underlying asset price. Under the Black–Scholes assumption, the delta of a call is defined as

$$\text{delta} = \frac{\partial c}{\partial S} = \Phi(d_1). \tag{7.10}$$

We use simulation to calculate the hedge ratio, delta, for general European options.

The risk-neutral valuation asserts that an option with payoff $F(S_T)$ can be valued as $e^{-rT}\hat{E}[F(S_T)|S_0 = S]$. Therefore, delta equals

$$\text{delta} = e^{-rT}\frac{\partial}{\partial S}\hat{E}[F(S_T)|S_0 = S]. \tag{7.11}$$

In order to compute delta under the Black–Scholes dynamics, the following theorem is established.

Theorem 7.3   *The delta of a European option with payoff $F(S_T)$ is given by*

$$\text{delta} = e^{-rT}\hat{E}\left[F(S_T)\frac{W_T}{S\sigma T}\right], \tag{7.12}$$

where W_T is the standard Brownian motion driving S_T.

Proof. Ignoring the discount factor, the definition of delta in Equation 7.11 is

$$\frac{\partial}{\partial S} \int_{-\infty}^{\infty} F(e^x)\phi(x| \log S) \, dx = \int_{-\infty}^{\infty} F(e^x) \frac{1}{S} \frac{\partial \phi(x| \log S)}{\partial \log S} \, dx,$$

where

$$\phi(x|y) = \frac{1}{\sigma\sqrt{2\pi T}} \exp\left[-\frac{(x-y-vT)^2}{2\sigma^2 T}\right].$$

Standard differentiation shows that

$$\frac{\partial \phi(x|y)}{\partial y} = \phi(x|y) \frac{x-y-vT}{\sigma^2 T}.$$

Hence, we have

$$\text{delta} = e^{-rT} \int_{-\infty}^{\infty} F(e^x) \frac{x - \log S - vT}{S\sigma^2 T} \phi(x| \log S) \, dx.$$

Recall that $x = \log S_T$,

$$x - \log S - vT = \log S_T - \log S - vT = \sigma W_T.$$

This completes the proof.   □

Theorem 7.3 enables us to simulate option delta (or even gamma) as follows.

Step 1: Generate $Z_1, Z_2, \dots, Z_n \sim N(0, 1)$ i.i.d.

Step 2: Set $Y_j = F\left(Se^{(r-\sigma^2/2)T+\sigma Z_j \sqrt{T}}\right) \frac{Z_j}{S\sigma\sqrt{T}}$.

Step 3: Set delta $= \frac{1}{n} \sum_{j=1}^{n} Y_j$.

The theorem can be extended to the case of path-dependent options. However, the derivation requires some knowledge of Malliavin calculus, which is beyond of the scope of the book. For details of this generalization, we refer to the article of Fournie et al. (1999, 2000).

Example 7.4   *The current price is $10, interest rate 5%, and volatility 40%. Simu-late the price and delta of a call option with strike price $12 and maturity 1 year by generating 10,000 terminal asset prices.*

An algorithm can be constructed as follows:

Step 1: Generate 10,000 terminal asset prices by the formula

$$S_T^j = S_0 \exp\left[(r-\sigma^2/2)T + \sigma\sqrt{T}Z_j\right], \quad Z_j \sim N(0, 1).$$

Step 2: For $j = 1$ to 10,000, Compute

$$C_j = \max(S_T^j - K) * \exp(-rT) \text{ and } Del_j = C_j * Z_j/(\sigma\sqrt{T}S_0).$$

Step 3: Compute call price = $\text{mean}(C_j)$ and delta = $\text{mean}(Del_j)$.

To simulate call option price and its delta in Visual Basic for Applications, go to the Online Supplementary and download the file *Chapter 7 Example 7.4 Simulating price and delta of a call option*.

With a CPU time of 0.9 s, our simulation finds that the call price is 1.06 and the delta is 0.44. The Black–Scholes call price and the delta are 1.08 and 0.448, respectively. This demonstrates the efficiency and accuracy of the simulation algorithm.

One thing we have to stress is that Theorem 7.3 is very useful for simulating deltas of single asset European options, with arbitrary payoff $F(S_T)$. However, it may not be applicable for path-dependent options and multiasset options. Therefore, we introduce alternative methods in later chapters.

7.4   EXERCISES

1. Write the VBA code for the newspaper boy example of Section 7.2.

2. Suppose that the asset return follows the t-distribution with two degrees of free-dom. Write a VBA code to simulate the 95% confidence VaR with parameters given in Example 7.1. Compare your result with the one obtained by normal VaR.

3. Implement the rejection method for generating GED when $v = 1.4$.

4. Verify Equations 7.5 and 7.6.

5. Verify Equations 7.7 and 7.8.

6. Let $S_0 = 100, \mu = r = 0.05, \sigma = 0.3$. Use the geometric Brownian motion method to simulate 20 daily prices of the stock $S_i, i = 1, \dots, 20$.

   (a) Suppose that you want to determine the price of a European put option maturing in 20 days with a strike price $K = 100$. Use simulation techniques to estimate this price.

   (b) Compare your result with the one obtained from the Black–Scholes formula. Are they similar?

7. The gamma of an option is defined as

$$\frac{\partial(\text{delta})}{\partial S}.$$

   (a) What is the financial interpretation of the gamma?

(b) By modifying the proof of Theorem 7.3, show that

$$\text{gamma} = e^{-rT}\hat{\mathrm{E}}\left[F(S_T)\frac{1}{S^2\sigma T}\left(\frac{W_T^2}{\sigma T} - W_T - \frac{1}{\sigma}\right)\right].$$

(c) Construct and implement a simulation algorithm to compute the call option gamma with a SPLUS code or VBA code.

(d) Suppose that $S = 10, K = 12, r = 0.1, \sigma = 0.3$, and $T = 0.8$. Compare your simulation result with the closed form solution:

$$\text{gamma} = \frac{1}{S\sigma\sqrt{2\pi T}}\exp\left[-\frac{\left(\ln\frac{S}{K} + (r + \sigma^2/2)T\right)^2}{2\sigma^2 T}\right].$$

The solutions and/or additional exercises are available online at http://www.sta. cuhk.edu.hk/Book/SRMS/.

7.5 APPENDIX

The data comprise 2,264 daily rates of returns. These data are transformed into standardized returns by using the sample mean and standard derivation. We assume that standardized returns follow a GED distribution with parameter ξ. Our goal is to estimate ξ. The density function of the GED distribution is given in Equation 7.4. Hence,

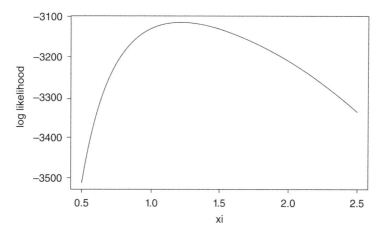

Figure 7.7 The log likelihood against ξ.

the likelihood function is

$$L(\xi) = \prod_{i=1}^{2,264} \frac{\xi \exp\left(-\frac{1}{2}\left|Z_i/\lambda\right|^{\xi}\right)}{\lambda 2^{1+1/\xi}\Gamma(1/\xi)},$$

$$\lambda = \left[\frac{2^{-2/\xi}\Gamma(1/\xi)}{\Gamma(3/\xi)}\right]^{1/2},$$

where $Z_1, \ldots, Z_{2,264}$ are standardized returns. Instead of deriving the MLE theoretically, we search the maximum point of the likelihood function with a numerical method. To confine the target point in a small interval, we plot the likelihood function against the parameter ξ. In Figure 7.7, we recognize that a unique maximum appears for $\xi \in (1, 1.3)$. The plot is given after this section. We then use the bisection method to search for the solution. Specifically, we compare $L(1)$ and $L(1.3)$ and discard the smaller one. The next step compares the remaining one with $L(1.15)$, the functional value at the mid-point of 1 and 1.3. We discard the point with a smaller value in L and repeat the procedure until a sufficiently accurate solution is obtained. Ultimately, $\xi = 1.21$, which has been input to generate GED-VaR in Section 7.2.3.

8

VARIANCE REDUCTION TECHNIQUES

8.1 INTRODUCTION

In standard Monte Carlo, we estimate the unknown quantity $\theta = \mathrm{E}X$ by generating random numbers $X_1, \ldots, X_n$ and use $\overline{X}_n$ to estimate θ. Recall that in the preceding chapter, the standard error for $\overline{X}_n$ is $\sigma/\sqrt{n}$, where σ^2 is the variance of X. There are two sources of contributions to the standard error of estimation. One is the factor $1/\sqrt{n}$, which is intrinsic to the Monte Carlo method, and not much can be done about it. The other one is the standard error σ of the output X, which by some techniques, can be improved upon. There are usually four standard methods to reduce σ:

1. Antithetic variables
2. Control variates
3. Stratification
4. Importance sampling

We discuss each of these methods in the subsequent sections.

8.2 ANTITHETIC VARIABLES

The idea of antithetic variables can best be illustrated by considering a special example. Suppose that we want to estimate $\theta = \mathrm{E}X$ by generating two outputs X_1

Simulation Techniques in Financial Risk Management, Second Edition. Ngai Hang Chan and Hoi Ying Wong.
© 2015 John Wiley & Sons, Inc. Published 2015 by John Wiley & Sons, Inc.

and X_2 such that $EX_1 = EX_2 = \theta$ and $\text{Var}X_1 = \text{Var}X_2 = \sigma^2$. Then

$$\text{Var}(\tfrac{1}{2}(X_1 + X_2)) = \tfrac{1}{4}(\text{Var}X_1 + \text{Var}X_2 + 2\text{Cov}(X_1, X_2))$$

$$= \frac{\sigma^2}{2} + \frac{1}{2}\text{Cov}(X_1, X_2)$$

$$\leq \frac{\sigma^2}{2}, \quad \text{if } \text{Cov}(X_1, X_2) \leq 0.$$

Note that when X_1 and X_2 are independent, then $\text{Var}((X_1 + X_2)/2) = \sigma^2/2$. Thus, the aforementioned inequality asserts that if X_1 and X_2 are negatively correlated, then the variance of the mean of the two would be less than the case when X_1 and X_2 were independent.

How do we generate negatively correlated random numbers? Suppose that we simulate $U_1, \ldots, U_m$, which are uniform random numbers. Then $V_1 = 1 - U_1, \ldots, V_m = 1 - U_m$ would also be uniform random numbers with the property that (U_i, V_i) being negatively correlated (exercise). If $X_1 = h(U_1, \ldots, U_m)$, then $X_2 = h(V_1, \ldots, V_m)$ must have the same distribution as X_1. It turns out that if h is a monotone function (either increasing or decreasing) in each of its arguments, then X_1 and X_2 are negatively correlated. This result is proved later at the end of this section. Thus, after generating $U_1, \ldots, U_m$ to compute X_1, instead of generating another new independent set of Us to compute X_2, we compute X_2 by

$$X_2 = h(V_1, \ldots, V_m) = h(1 - U_1, \ldots, 1 - U_m).$$

Accordingly, $(X_1 + X_2)/2$ should have smaller variance.

In general, we may generate $X_i = F^{-1}(U_i)$ using the inverse transform method. Let $Y_i = F^{-1}(V_i)$. As F is monotone, so is F^{-1} and, hence, X_i and Y_i will be negatively correlated. Both $X_1, \ldots, X_n$ and $Y_1, \ldots, Y_n$ generated in this way are i.i.d. (identical and independent distributed) sequences with c.d.f. (cumulative distribution function) F, but negatively correlated.

Definition 8.1   *The Y_i sequence is called the sequence of antithetic variables.*

For normal distributions, generating antithetic variables is straightforward. Suppose that $X_i \sim N(\mu, \sigma^2)$, then $Y_i = 2\mu - X_i$ also has a normal distribution with mean μ and variance σ^2 and X_i and Y_i are negatively correlated.

More generally, if we want to compute $E(H(X))$ for some function H, standard Monte Carlo suggests using $\frac{1}{n}\sum_{i=1}^n H(X_i)$. Then an antithetic estimator of $E(H(X))$ is

$$\hat{H}_{AN} = \frac{1}{2n}\sum_{i=1}^n (H(X_i) + H(Y_i)),$$

where Y_i is a sequence of antithetic variables. To see how variance reduction is achieved by using this antithetic estimator, let $\text{Var}(H(X)) = \sigma^2$ and

$\mathrm{Corr}(H(X), H(Y)) = \rho$. Consider

$$\mathrm{Var}(\hat{H}_{AN}) = \frac{1}{4n^2} \sum_{i=1}^{n} \{\mathrm{Var}H(X_i) + \mathrm{Var}H(Y_i) + 2\mathrm{Cov}(H(X_i), H(Y_i))\}$$

$$= \frac{1}{4n^2}(2n\sigma^2 + 2n\rho\sigma^2)$$

$$= \frac{\sigma^2}{2n}(1 + \rho).$$

Note that when $H(X)$ and $H(Y)$ are uncorrelated ($\rho = 0$), then the variance would be reduced by a factor of 2, which is equivalent to doubling the simulation size. On the other hand, if $\rho = -1$, then the variance would be reduced to zero. As long as ρ is negative, some form of variance reduction can be achieved. An obvious question is that in view of this observation, why not choose Y so that $\rho = -1$? Such Ys may be difficult to construct, as ρ represents the correlation between $H(X)$ and $H(Y)$. In the case $H(X) = X$, then $\hat{H}_{AN}$ reduces to a constant, which is the perfect scenario. In view of these caveats, we usually choose the antithetic variables Y so that ρ is negative, not necessarily -1. When H is linear, such as the case $H(X) = X$, the antithetic variable works best. In general, the more linear the H is, the more effective the antithetic variable will be.

Example 8.1   *Let $\theta = \mathrm{E}(e^U) = \int_0^1 e^x\, dx$.*

We know that $\theta = e - 1$. Consider the antithetic variable $V = 1 - U$. Recall that the moment-generating function of U equals $\mathrm{E}(e^{tU}) = (e^t - 1)/t$. Now

$$\mathrm{Cov}(e^U, e^V) = \mathrm{E}(e^U e^V) - \mathrm{E}(e^U)\mathrm{E}(e^V)$$

$$= \mathrm{E}(e^U e^{1-U}) - \mathrm{E}(e^U)\mathrm{E}(e^{1-U})$$

$$= e - (e - 1)^2 = -0.2342.$$

Furthermore,

$$\mathrm{Var}(e^U) = \mathrm{E}(e^{2U}) - (\mathrm{E}(e^U))^2 = (e^2 - 1)/2 - (e - 1)^2 = 0.242.$$

Thus, for U_1 and U_2 to be independent uniform $(0, 1)$ random variables,

$$\mathrm{Var}[(e^{U_1} + e^{U_2})/2] = \mathrm{Var}(e^U)/2 = 0.121.$$

But

$$\mathrm{Var}[(e^U + e^V)/2] = \mathrm{Var}(e^U)/2 + \mathrm{Cov}(e^U, e^V)/2 = 0.121 - 0.2342/2 = 0.0039,$$

achieving a substantial variance reduction of 96.7%.

We are now ready to justify the argument used in advocating antithetic variables.

Theorem 8.1   *Let $X_1, \ldots, X_n$ be independent, then for any increasing functions f and g of n variables,*

$$E(f(X)g(X)) \geq Ef(X)Eg(X),$$

where $X = (X_1, \ldots, X_n)$.

Proof. By mathematical induction. Consider $n = 1$, then

$$(f(x) - f(y))(g(x) - g(y)) \geq 0, \text{ for all } x \text{ and } y,$$

as both factors are either non-negative $(x \geq y)$ or nonpositive $(x \leq y)$. Thus, for any random variables X and Y,

$$(f(X) - f(Y))(g(X) - g(Y)) \geq 0 \quad \text{implying} \quad E((f(X) - f(Y))(g(X) - g(Y))) \geq 0.$$

In other words,

$$E(f(X)g(X)) + E(f(Y)g(Y)) \geq E(f(X)g(Y)) + E(f(Y)g(X)).$$

If X and Y are independent and identically distributed, then

$$E(f(X)g(X)) = E(f(Y)g(Y))$$

and

$$E(f(X)g(Y)) = E(f(Y)g(X)) = E(f(Y))E(g(X)) = E(f(X))E(g(X))$$

so that

$$E(f(X)g(X)) \geq E(f(X))E(g(X)),$$

proving the result for the case $n = 1$. Assume the result for $n - 1$. Suppose $X_1, \ldots, X_n$ are independent and let f and g be increasing functions. Then

$$E(f(X)g(X)|X_n = x_n) = E(f(X_1, \ldots, X_{n-1}, x_n)g(X_1, \ldots, X_{n-1}, x_n)|X_n = x_n)$$

$$= E(f(X_1, \ldots, X_{n-1}, x_n)g(X_1, \ldots, X_{n-1}, x_n))$$

$$\text{(because of independence)}$$

$$\geq E(f(X_1, \ldots, X_{n-1}, x_n))E(g(X_1, \ldots, X_{n-1}, x_n))$$

$$\text{(by induction hypothesis)}$$

$$= E(f(X)|X_n = x_n)E(g(X)|X_n = x_n).$$

Hence,

$$E(f(X)g(X)|X_n) \geq E(f(X)|X_n)E(g(X)|X_n).$$

On taking expectation on both sides of this equation, we have

$$E(f(X)g(X)) \geq E[E(f(X)|X_n)E(g(X)|X_n)].$$

Observe that $E(f(X)|X_n)$ and $E(g(X)|X_n)$ are increasing functions of X_n so that by the result of $n = 1$, we have

$$E[E(f(X)|X_n)E(g(X)|X_n)] \geq E[E(f(X|X_n))]E[E(g(X|X_n))]$$
$$= E(f(X))E(g(X)).$$

This completes the proof for the case of n.   □

Corollary 8.1   *If $h(X_1, \ldots, X_n)$ is a monotone function of each of its arguments, then for a set $U_1, \ldots, U_n$ of independent random numbers,*

$$\mathrm{Cov}[h(U_1, \ldots, U_n), h(1 - U_1, \ldots, 1 - U_n)] \leq 0.$$

Proof. Without loss of generality, by redefining h, we may assume that h is increasing in its first r arguments and decreasing in its remaining $n - r$ arguments. Let

$$f(x_1, \ldots, x_n) = h(x_1, \ldots, x_r, 1 - x_{r+1}, \ldots, 1 - x_n),$$
$$g(x_1, \ldots, x_n) = -h(1 - x_1, \ldots, 1 - x_r, x_{r+1}, \ldots, x_n).$$

It follows that both f and g are increasing functions. By the preceding theorem,

$$\mathrm{Cov}[f(U_1, \ldots, U_n), g(U_1, \ldots, U_n)] \geq 0.$$

That is,

$$\mathrm{Cov}[h(U_1, \ldots, U_r, V_{r+1}, \ldots, V_n), h(V_1, \ldots, V_r, U_{r+1}, \ldots, U_n)] \leq 0, \qquad (8.1)$$

where $V_i = 1 - U_i$. Observe that as $(h(U_1, \ldots, U_n), h(V_1, \ldots, V_n))$ has the same joint distribution as $(h(U_1, \ldots, U_r, V_{r+1}, \ldots, V_n), h(V_1, \ldots, V_r, U_{r+1}, \ldots, U_n))$, it follows from Equation 8.1 that

$$\mathrm{Cov}[h(U_1, \ldots, U_n), h(V_1, \ldots, V_n)] \leq 0,$$

proving the corollary.   □

When is antithetic variable effective? The following are some guidelines:

- Antithetic variables will result in a lower variance estimate than independent simulations only if the values computed from a path and its antithetic variables are *negatively correlated*.

- If H is monotone in each of its arguments, then antithetic variables reduce variance in estimating $E(H(Z_1, \ldots, Z_n))$.
- If H is linear, then an antithetic estimate of $E(H(Z_1, \ldots, Z_n))$ has zero variance.
- If H is symmetric, that is, $H(-Z) = H(Z)$, then an antithetic estimate of sample size $2n$ has the same variance as an independent sample of size n.

Example 8.2 *To illustrate some of these points, consider the simulations of payoff of options using antithetic variables. The function H in this case maps*

$$z \to \max\{0, S_0 \exp([r - \sigma^2/2]T + \sigma\sqrt{T}z) - K\}.$$

In Figure 8.1, the vertical axis is the payoff and the horizontal axis is the value of z, the input standard normal. All cases have $r = 0.05\%$, $K = 50$, and $T = 0.5$. The top three cases have $\sigma = 0.3$ and S_0=40, 50, and 60; the second three cases have $S_0 = 50$ and $\sigma = 0.10, 0.20, 0.30$. The top three graphs correspond to the function H for options that are out-of-money ($S_0 = 40$), at-the-money ($S_0 = 50$), and in-the-money ($S_0 = 60$), respectively; the bottom three graphs correspond to low, intermediate, and high volatility for an at-the-money option. (The precise parameter values are given in the caption of the figure.) As one would expect, increasing moneyness and decreasing volatility both increase the degree of linearity. For the values indicated in the figure, we find numerically that antithetics reduce variance by $14\%, 42\%$, and 80% in the top three cases and by $65\%, 49\%$, and 42% in the bottom three, respectively. Clearly, the more linear the function H is, the more effective the antithetic variable technique will be.

Example 8.3 *Figure 8.2 plots the payoff of $|S_T - K|$ on a straddle as a function of z. The parameter values are given in the caption. The graph shows a high degree of symmetry around zero, suggesting that antithetic variables may not be as effective as in the other cases. Numerical results here indicate that an antithetic estimate based on m pairs of antithetic variables has higher variance than an estimate based on $2m$ independent samples.*

Please see the online material for the VBA codes.

8.3 STRATIFIED SAMPLING

The idea of stratification is often used in sample surveys (Barnett, 1991). The idea lies in the observation that the population may be heterogeneous and consists of various homogeneous subgroups (such as gender, race, and social–economic status). If we wish to learn about the whole population (such as whether people in Hong Kong would like to have universal suffrage in 2007), we can take a random sample from the whole population to estimate that quantity. On the other hand, it would be more efficient to take small samples from each subgroup and combine the estimates in each subgroup according to the fraction of the population that subgroup represents. As

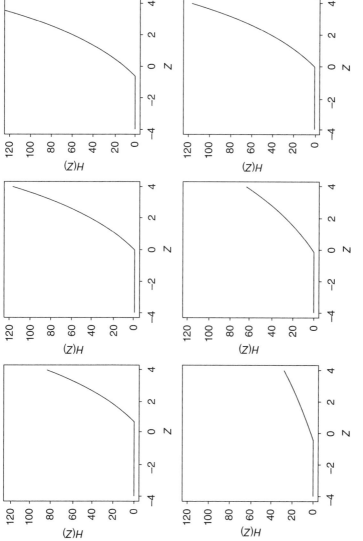

Figure 8.1 Illustration of payoffs for antithetic comparisons.

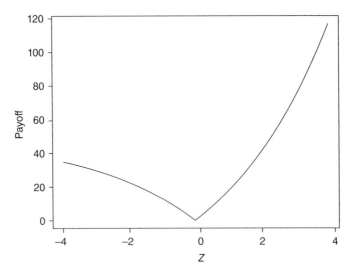

Figure 8.2 Payoff on a straddle as a function of input normal Z based on the parameters $S_0 = K = 50$, $\sigma = 0.30$, $T = 1$, and $r = 0.05$.

we can learn about the opinion of a homogeneous subgroup with a relatively small sample size, this stratified sampling procedure would be more efficient.

In general, if we want to estimate EX, where X depends on a random variable S that takes on one of the values in $\{1, \ldots, k\}$ with known probabilities, then the technique of stratification runs into k groups, with the ith group having $S = i$. Let $\overline{X}_i$ be the average values of X in those runs having $S = i$, and then estimate $EX = \sum_{i=1}^{k} E(X|S = i)P(S = i)$ by

$$\sum_{i=1}^{k} \overline{X}_i P(S = i).$$

This is known as stratified sampling.

To illustrate this idea, suppose that we want to estimate $E(g(U)) = \int_0^1 g(x)dx$. Consider two estimators on the basis of a sample of $2n$ runs. The first one is the standard method,

$$\hat{g} = \frac{1}{2n} \sum_{i=1}^{2n} g(U_i).$$

Note that $E(\hat{g}) = E(g(U))$ and

$$\text{Var}(\hat{g}) = \frac{1}{4n^2} \sum_{i=1}^{2n} \text{Var}(g(U_i)) = \frac{1}{2n} \left[\int_0^1 g^2(x)\,dx - \left(\int_0^1 g(x)\,dx \right)^2 \right].$$

On the other hand, we can write

$$E(g(U)) = \int_0^{1/2} g(x)\,dx + \int_{1/2}^1 g(x)\,dx.$$

Instead of selecting Us from $[0, 1]$, we can select the first n Us from $[0, 1/2]$ and the remaining n Us from $[1/2, 1]$ to construct a new estimator

$$\hat{g}_s = \frac{1}{2n}\left[\sum_{i=1}^n g(U_i/2) + \sum_{i=n+1}^{2n} g((U_i + 1)/2)\right].$$

It can be easily seen that if $U \sim U(0, 1)$, then $V = a + (b - a)U$ is distributed as uniform (a, b). In particular, $U/2 \sim U(0, 1/2)$ and $(U + 1)/2 \sim U(1/2, 1)$. To compute the variance of the new estimator, consider

$$\text{Var}(\hat{g}_s) = \frac{1}{4n^2}\left\{\sum_{i=1}^n \text{Var}(g(U_i/2)) + \sum_{i=n+1}^{2n} \text{Var}(g((U_i + 1)/2))\right\}.$$

Direct computations show that if $U_i \sim U(0, 1)$, then

$$\text{Var}\left(g\left(\frac{U_i}{2}\right)\right) = 2\int_0^{1/2} g^2(x)\,dx - 4m_1^2,$$

$$\text{Var}\left(g\left(\frac{U_i + 1}{2}\right)\right) = 2\int_{1/2}^1 g^2(x)\,dx - 4m_2^2,$$

where $m_1 = \int_0^{1/2} g(x)\,dx$ and $m_2 = \int_{1/2}^1 g(x)\,dx$. Now

$$\text{Var}\left(g\left(\frac{U_i}{2}\right)\right) + \text{Var}\left(g\left(\frac{U_i + 1}{2}\right)\right) = 2\int_0^1 g^2(x)\,dx - 4(m_1^2 + m_2^2).$$

Consequently,

$$\text{Var}(\hat{g}_s) = \frac{1}{2n}\left\{\int_0^1 g^2(x)\,dx - 2(m_1^2 + m_2^2)\right\}.$$

Note that

$$(m_1 + m_2)^2 + (m_1 - m_2)^2 = 2(m_1^2 + m_2^2).$$

Therefore,

$$\text{Var}(\hat{g}_s) = \frac{1}{2n}\left\{\int_0^1 g^2(x)\,dx - (m_1 + m_2)^2 - (m_1 - m_2)^2\right\}$$

$$= \text{Var}(\hat{g}) - \frac{1}{2n}(m_1 - m_2)^2.$$

Because this second term is always non-negative, stratification reduces the variance by an amount of this second term. The bigger the difference in m_1 and m_2, the greater the reduction in variance. In general, if more strata are introduced, more reduction will be achieved. One can generalize this result to the multistrata case, but we omit the mathematical details here.

Example 8.4   *Consider again* $\theta = E(e^U) = \int_0^1 e^x \, dx$.

Recall that by standard Monte Carlo with $n = 2$,

$$\hat{g} = \frac{1}{2}(e^{U_1} + e^{U_2}),$$

and $\mathrm{Var}(\hat{g}) = 0.121$. On the other hand, using stratification, we have

$$\hat{g}_s = \frac{1}{2}(e^{U_1/2} + e^{(U_2+1)/2}),$$

and $\mathrm{Var}(\hat{g}_s) = \mathrm{Var}(\hat{g}) - (m_1 - m_2)^2/2$,   where $m_1 = \int_0^{1/2} e^x \, dx = e^{1/2} - 1$ and $m_2 = \int_{1/2}^1 e^x \, dx = e - e^{1/2}$. Thus,

$$\mathrm{Var}(\hat{g}_s) = 0.121 - (2e^{1/2} - e - 1)^2/2 = 0.0325,$$

resulting a variance reduction of 73.13%.

Stratified sampling is also very useful to draw random samples from designated ranges. For example, if we want to sample $Z_1, \ldots, Z_{100}$ from a standard normal distribution, the standard technique would partition the whole real line $(-\infty, \infty)$ into a number of bins and sample Zs from these bins randomly. In such a case, it is inevitable that some bins may have more samples, while other bins, particularly those near the tails, may have no sample at all. Therefore, a random sample drawn this way would under-represent the tails. Although this may not be a serious issue in general, it may have severe effect when the tail is the quantity of interest, such as the case in the simulation of VaR. To ensure that the bins are regularly represented, we may generate the Zs as follows. Let

$$V_i = \frac{1}{100}(U_i + (i - 1)), \quad i = 1, \ldots, 100,$$

where $U_i \sim U(0,1)$ i.i.d.. By the property of uniform distribution, $V_i \sim U(\frac{i-1}{100}, \frac{i}{100})$. Now let $Z_i = \Phi^{-1}(V_i)$. Then Z_i falls between the $i - 1$ and i percentiles of the standard normal distribution. For example, if $i = 1$, then $V = U/100 \sim U(0, 1/100)$ so that $Z = \Phi^{-1}(V)$ falls between $\Phi^{-1}(0) = -\infty$ and $\Phi^{-1}(0.01)$, that is, the 0th and the 1st percentile of a standard normal distribution.

This method gives equal weight to each of the 100 equiprobable strata. Of course, the number 100 can be replaced by any number that is desirable. The price we pay in stratification is the loss of independence of the Zs. This complicates statistical inference for simulation results.

Example 8.5 *As an illustration of stratification, consider simulating standard normal random numbers via standard method and stratification method, respectively. As can be clearly seen from Figures 8.3 and 8.4, stratified sampling generates samples much more uniformly over the range than standard Monte Carlo. Please see the online material for the VBA codes.*

Example 8.6 *As a second illustration of stratification, consider the simulation of a European call option of Example 7.2 again.*

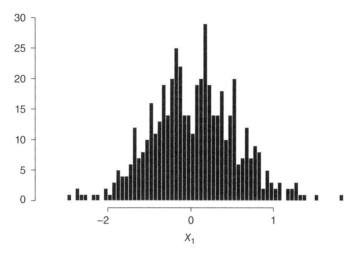

Figure 8.3   Simulations of 500 standard normal random numbers by standard Monte Carlo.

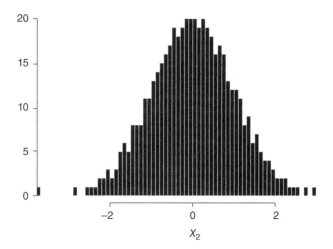

Figure 8.4   Simulations of 500 standard normal random numbers by stratified sampling.

In Example 7.2, we simulate the terminal prices $S_1(T), \ldots, S_n(T)$ according to Equation 5.1 and then compute the estimate as

$$\overline{C} = \frac{e^{-rT}}{n} \sum_{i=1}^{n} \max\{S_i(T) - K, 0\}.$$

In this standard simulation, the random normals are samples arbitrarily over the whole real line. We can improve the efficiency by introducing stratification.

1. Partition $(-\infty, \infty)$ into B strata or bins.
2. Set $V_i = \frac{1}{B}(U_i + (i - 1))$, $i = 0, \ldots, B$ and generate the desired number of random samples (N_B, say) of Vs in the ith bin.
3. Apply $\Phi^{-1}(V_i)$ to get the desired normal random numbers from each bin and calculate $\overline{C}_i$ from each bin.
4. Average the $\overline{C}_i$ over the total number of bins to get an overall estimate $\overline{C}$.
5. Calculate the standard error as in the previous cases.

This numerical example uses $S_0 = 10$, $K = 12$, $r = 0.03$, $\sigma = 0.40$, and $T = 1$. The theoretical Black–Scholes price is 1.0139. We simulate the European option price for different bin sizes with $N_B \times B = 1,000$ in all cases. The effect of stratification increases as we increase the number of bins. The results are shown in Table 8.1. Please see the online material for the VBA codes.

Regular stratification puts equal weight on each of the B bins. Such an allocation may not be ideal as one would like to have sample sizes directly related to the variability of the target function over that bin. To illustrate this point, consider the payoff of a European call option again.

Example 8.7 *Stratified sampling for a European call with the same parameter values as in Example 8.6.*

TABLE 8.1   Effects of Stratification for Simulated Option Prices with Different Bin Sizes

Bins (B)	N_B	Mean ($\overline{C}$)	Std. Err.
1	1000	0.9744	0.0758
2	500	1.0503	0.0736
5	200	1.0375	0.0505
10	100	0.9960	0.0389
20	50	0.9874	0.0229
50	20	1.0168	0.0146
100	10	0.9957	0.0092
200	5	1.0208	0.0094
500	2	1.0151	0.0062
1000	1	1.0091	NA

We know that if $S_T < K$, then the payoff of the call is zero. Recall that

$$S_T = S_0 e^{[r - \sigma^2/2]T + \sigma\sqrt{T}Z}.$$

Therefore, $S_T < K$ if $S_0 e^{[r - \sigma^2/2]T + \sigma\sqrt{T}Z} < K$. That is,

$$Z < [\log(K/S_0) - (r - \sigma^2/2)T]/(\sigma\sqrt{T}) := L.$$

Every simulated $Z < L$ is being wasted as it just returns the value 0. We should only be concentrating on the interval $[L, \infty)$. How can we achieve this goal?

1. Find out the c.d.f. of a normal distribution Y restricted on $[L, \infty)$. It can be shown that Y has c.d.f.
   $$F(y) = \frac{\Phi(y) - \Phi(L)}{1 - \Phi(L)}.$$

2. Use the inverse transform method to generate Y. Consider the inverse transformation of F, that is, solve for y such that $y = F^{-1}(x)$. Writing it out, we have $x = F(y) = \frac{\Phi(y) - \Phi(L)}{1 - \Phi(L)}$ so that

   $$y = \Phi^{-1}(x(1 - \Phi(L)) + \Phi(L)).$$

   Now generate U from uniform $(0, 1)$ and evaluate

   $$Y = \Phi^{-1}(U(1 - \Phi(L)) + \Phi(L)).$$

3. Plug in the generated Y into the simulation step of the payoff of the call and complete the analysis. Note that when evaluating the new estimator for the payoff, we need to multiply the factor $1 - \Phi(L)$. That is,

   $$C^* = (1 - \Phi(L))\overline{C},$$

   where $\overline{C}$ is the average of the simulated payoffs using the truncated normal random variables.

In general, we would like to apply the stratification technique to bins in which the variability of the integrand is largest. In this case, we just focus the entire sample on the case $S_T > K$.

The results are given in Table 8.2. Please see the online material for the VBA codes.

TABLE 8.2 Effects of Stratification for Simulated Option Prices with Restricted Normal

Bins (B)	N_B	Mean $(\overline{C})$	Std. Err.	Adj. Mean	SE
1	1000	0.9744	0.0758	0.9842	0.1102
2	500	1.0503	0.0736	1.0303	0.0823
5	200	1.0375	0.0505	1.0235	0.0524
10	100	0.9960	0.0389	1.0101	0.0404
20	50	0.9874	0.0229	1.0058	0.0238
50	20	1.0168	0.0146	1.0147	0.0153
100	10	0.9957	0.0092	1.0089	0.0095
200	5	1.0208	0.0094	1.0160	0.0099
500	2	1.0151	0.0062	1.0143	0.0066
1000	1	1.0091	NA	1.0125	NA

8.4 CONTROL VARIATES

The idea of control variates is very simple. Suppose that we want to estimate $\theta = EX$ from the simulated data. Suppose that for some other variable Y, the mean $\mu_Y = EY$ is known. Then for any given constant c, the quantity

$$X_{CV} = X + c(Y - \mu_Y)$$

is also an unbiased estimate of θ, as $E(X_{CV}) = \theta$. Presumably, if we choose the constant c cleverly, some form of variance reduction can be achieved. How can we do this? In other words, what would be a good choice of c? To answer this question, first consider the variance of the new estimator X_{CV}, call it σ^2_{CV}.

$$\sigma^2_{CV} = \text{Var}(X + c(Y - \mu_Y)) = \text{Var}X + c^2\text{Var}Y + 2c\text{Cov}(X, Y).$$

We would like to find c such that σ^2_{CV} is minimized. Differentiate the preceding expression with respect to c and set it equal to zero, we have

$$2c\text{Var}Y + 2\text{Cov}(X, Y) = 0.$$

Solving for such a c, we get, $c^* = -\text{Cov}(X, Y)/\text{Var}Y$ as the value of c that minimizes σ^2_{CV}. For such a c^*,

$$\sigma^2_{c^*} = \text{Var}X - \frac{\text{Cov}^2(X, Y)}{\text{Var}Y}.$$

The variable Y used in this way is known as a control variate for the simulation estimator X. Recall that $\text{Corr}(X, Y) = \text{Cov}(X, Y)/(\text{Var}X\text{Var}Y)^{1/2}$. Therefore,

$$\sigma^2_{c^*} = \text{Var}X(1 - \text{Corr}^2(X, Y)).$$

Hence, as long as $\mathrm{Corr}(X, Y) \neq 0$, some form of variance reduction is achieved. In practice, quantities such as $\sigma_Y^2 = \mathrm{Var}\, Y$ and $\mathrm{Cov}(X, Y)$ are usually not available; they have to be estimated from the simulations on the basis of sample values. For example, let $\overline{X} = \sum_{i=1}^{n} X_i / n$ and $\overline{Y} = \sum_{i=1}^{n} Y_i / n$. Then

$$\hat{\mathrm{Cov}}(X, Y) = \frac{1}{n-1} \sum_{i=1}^{n} (X_i - \overline{X})(Y_i - \overline{Y}),$$

$$\hat{\sigma}_Y^2 = \frac{1}{n-1} \sum_{i=1}^{n} (Y_i - \overline{Y})^2,$$

$$\hat{c}^* = -\frac{\hat{\mathrm{Cov}}(X, Y)}{\hat{\sigma}_Y^2}.$$

Suppose that we use $\overline{X}$ from simulation to estimate θ. Then the control variate would be $\overline{Y}$, and the control variate estimator is

$$\overline{X} + c^*(\overline{Y} - \mu_Y),$$

with variance equaling to

$$\frac{1}{n}\left(\mathrm{Var}\, X - \frac{\mathrm{Cov}^2(X, Y)}{\mathrm{Var}\, Y}\right) = \frac{\sigma_X^2}{n}(1 - \rho^2).$$

Equivalently, one can use the simple linear regression equation

$$X = a + bY + e, \quad e \sim \text{i.i.d. } (0, \sigma^2), \tag{8.2}$$

to estimate c^*. In fact, it can be easily shown that the least squares estimates of b, $\hat{b} = -\hat{c}^*$, see Weisberg (1985). In such a case, the control variate estimator is given by

$$\overline{X} + c^*(\overline{Y} - \mu_Y) = \overline{X} - \hat{b}(\overline{Y} - \mu_Y) = \hat{a} + \hat{b}\mu_Y, \tag{8.3}$$

where $\hat{a} = \overline{X} - \hat{b}\overline{Y}$ is the least squares estimate of a in Equation 8.2. That is, the control variate estimate is equal to the estimated regression equation evaluated at the point μ_Y.

Notice that there is a very simple geometric interpretation using Equation 8.2. Firstly, observe that the estimated regression line

$$\hat{X} = \hat{a} + \hat{b}Y$$
$$= \overline{X} + \hat{b}(Y - \overline{Y}).$$

Thus, this line passes through the point $(\overline{Y}, \overline{X})$. Secondly, from Equation 8.3,

$$\hat{X}_{CV} = \hat{a} + \hat{b}\mu_Y = \overline{X} - \hat{b}(\overline{Y} - \mu_Y).$$

Suppose that $\overline{Y} < \mu_Y$, that is, the simulation run underestimates μ_Y and suppose that X and Y are positively correlated. Then it is likely that $\overline{X}$ would underestimate $E(X) = \theta$. We therefore need to adjust the estimator upward, and this is indicated by the fact that $\hat{b} = -\hat{c}^* > 0$. The extra amount that needs to be adjusted upward equals $-\hat{b}(\overline{Y} - \mu_Y)$, which is governed by the linear Equation 8.3.

Finally, $\hat{\sigma}^2$, the regression estimate of σ^2, is the estimate of $\text{Var}(X - \hat{b}Y) = \text{Var}(X + \hat{c}^* Y)$. To see this, recall from regression that

$$\hat{\sigma}^2 = \frac{1}{n} \sum_{i=1}^{n} \hat{e}_i^2$$

$$= \frac{1}{n} \sum_{i=1}^{n} (X_i - \hat{a} - \hat{b} Y_i)^2$$

$$= \frac{1}{n} \sum_{i=1}^{n} (X_i - (\overline{X} - \hat{b}\overline{Y}) - \hat{b} Y_i)^2$$

$$= \frac{1}{n} \sum_{i=1}^{n} ((X_i - \overline{X}) - \hat{b}(Y_i - \overline{Y}))^2$$

$$= \frac{1}{n} \sum_{i=1}^{n} ((X_i - \overline{X})^2 - \hat{b}^2 (Y_i - \overline{Y})^2)$$

$$= \widehat{\text{Var}}(X) - \hat{b}^2 \widehat{\text{Var}}(Y)$$

$$= \widehat{\text{Var}}(X - \hat{b}Y).$$

The last equality follows from a standard expansion of the variance estimate (see exercise 8.2). It follows that the estimated variance of the control variate estimator $\overline{X} + \hat{c}^*(\overline{Y} - \mu_Y)$ is $\hat{\sigma}^2 / n$.

Example 8.8 *Consider the problem $\theta = E(e^U)$ again.*

Clearly, the control variate is U itself. Now

$$\text{Cov}(e^U, U) = E(Ue^U) - E(U)E(e^U)$$

$$= \int_0^1 xe^x \, dx - (e - 1)/2$$

$$= 1 - (e - 1)/2 = 0.14086.$$

The second last equality makes use of the facts from the previous examples that $E(U) = 1/2$, $\text{Var}\, U = 1/12$, and $\text{Var}(e^U) = 0.242$. It follows that the control variate estimate has variance

$$\text{Var}(e^U + c^*(U - 1/2)) = \text{Var}(e^U)(1 - 12(0.14086)^2 / 0.242) = 0.0039,$$

resulting in a variance reduction of $(0.242 - 0.0039)/0.242 \times 100\% = 98.4\%$.

In general, if we want to have more than one control variate, we can make use of outputs from the multiple linear regression model given by

$$X = a + \sum_{i=1}^{k} b_i Y_i + e, \quad e \sim \text{i.i.d. } (0, \sigma^2).$$

In this case, the least squares estimates of a and b_is, $\hat{a}$ and $\hat{b}_i$s can be easily shown to satisfy $\hat{c}_i^* = -\hat{b}_i$, $i = 1, \ldots, k$. Furthermore, the control variate estimate is given by

$$\overline{X} + \sum_{i=1}^{k} (\overline{Y}_i - \mu_i) = \hat{a} + \sum_{i=1}^{k} \hat{b}_i \mu_i,$$

where $E(Y_i) = \mu_i$, $i = 1, \ldots, k$. In other words, the control variate estimate is equal to the estimated multiple regression line evaluated at the point $(\mu_1, \ldots, \mu_k)$. By the same token, the variance of the control variate estimate is given by $\hat{\sigma}^2/n$, where $\hat{\sigma}^2$ is the regression estimate of σ^2.

Example 8.9 *Plunging along the same line, consider simulating the vanilla European call option as in Example 8.6, using the terminal value S_T as the control variate.*

The control variate estimator is given by

$$C_{CV} = C + c^*(S_T - E(S_T)).$$

Recalling that $S_T = S_0 e^{(\nu T + \sigma \sqrt{T} Z)}$, it can be easily deduced that

$$E(S_T) = S_0 e^{rT}, \tag{8.4}$$

$$\text{Var}(S_T) = S_0^2 e^{2rT}(e^{\sigma^2 T} - 1). \tag{8.5}$$

The algorithm goes as follows:

1. For $i = 1, \ldots, N_1$, simulate a pilot of N_1 independent paths to get

$$S_T(i) = S_0 e^{\nu T + \sigma \sqrt{T} Z_i},$$

$$C(i) = e^{-rT} \max\{0, S_T(i) - K\}.$$

2. Compute $E(S_T)$ as $S_0 e^{rT}$ or estimate it by $\sum_{i=1}^{N_1} S_T(i)/N_1$. Compute $\text{Var}(S_T)$ as $S_0^2 e^{2rT}(e^{\sigma^2 T} - 1)$ or estimate it by $\frac{1}{N_1 - 1} \sum_{i=1}^{N_1} (S_T(i) - \overline{S}_T)^2$. Now estimate covariance by

$$\hat{\text{Cov}}(S_T, C) = \frac{1}{N_1 - 1} \sum_{i=1}^{N_1} (S_T(i) - \overline{S}_T)(C(i) - \overline{C}),$$

where $\overline{C} = \sum_{i=1}^{N_1} C(i)/N_1$ and $\overline{S}_T = \sum_{i=1}^{N_1} S_T(i)/N_1$.

3. Repeat the simulations of S_T and C by means of control variate. For $i = 1, \ldots, N_2$, independently simulate

$$S_T(i) = S_0 e^{\nu T + \sigma \sqrt{T} Z_i},$$
$$C(i) = e^{-rT} \max\{0, S_T(i) - K\},$$
$$C_{CV}(i) = C(i) + c^*(S_T(i) - \mathrm{E}(S_T(i))),$$

where $c^* = -\widehat{\mathrm{Cov}}(S_T, C)/\widehat{\mathrm{Var}} S_T$ is computed from the preceding step.

4. Calculate the control variate estimator by

$$\overline{C}_{CV} = \frac{1}{N_2} \sum_{i=1}^{N_2} C_{CV}(i).$$

Complete the simulation by evaluating the standard error of $\overline{C}_{CV}$ and construct confidence intervals.

Please see the online material for the VBA codes.

For $N_1 = 500$ and $N_2 = 50,000$, we have a 95% confidence interval for C_{CV} of [1.0023 1.0247]. In this case, the estimated call price is 1.0135 with standard error 0.0057.

In using control variates, there are a number of features that should be considered.

- What should constitute the appropriate control? We have seen that in simple cases, the underlying asset prices may be appropriate. In more complicated situation, we may use some easily computed quantities that are highly correlated with the object of interest as control variates. For example, standard calls and puts frequently provide convenient source of control variates for pricing exotic options, and so does the underlying asset itself.
- The control variate estimator is usually unbiased by construction. In addition, we can separate the estimation of the coefficients ($\hat{c}_i^*$) from the estimation of prices.
- The flexibility of choosing the c_is suggests that we can sometimes make optimal use of information. In any event, we should exploit the specific feature of the problem under consideration, rather than generic applications of routine methods.
- Because of its close relationship with linear regression, control variates are easily computed and explained.
- We have only covered linear control. In practice, one can consider using nonlinear control variates, for example, $\overline{X}\,\overline{Y}/\mu_Y$. Statistical inference for nonlinear control may be tricky though.

8.5   IMPORTANCE SAMPLING

After studying three variance reduction methods, we pursue one last method, namely, importance sampling. This method is similar in idea to the acceptance–rejection method that was discussed in Chapter 6. Its main idea lies in approximating at places where the quantity of interest carries the most information, hence the name importance sampling. This chapter then concludes with examples illustrating the different methods of variance reduction in risk management.

Suppose that we are interested in estimating

$$\theta = E[h(X)] = \int h(x)f(x)\,dx,$$

where $X = (X_1, \dots, X_n)$ denotes an n-dimensional random vector having a joint p.d.f. $f(x) = f(x_1, \dots, x_n)$. Suppose that a direct simulation of the random vector X is inefficient so that computing $h(x)$ is infeasible. This inefficiency may be due to difficulties encountered in simulating X, or the variance of $h(x)$ being too large, or a combination of both.

Suppose that there exists another density $g(x)$, which is easy to simulate and satisfies the condition that $f(x) = 0$ whenever $g(x) = 0$. Then θ can be estimated by

$$\theta = E[h(x)]$$
$$= \int \frac{h(x)f(x)}{g(x)} g(x)\,dx$$
$$= E_g \left[\frac{h(x)f(x)}{g(x)} \right],$$

where the notation E_g denotes the expectation of the random vector X taken under the density g, that is, X has joint p.d.f. $g(x)$. It follows from this identity that θ can be estimated by generating X with density g and then using as the estimator the average of the values of $h(X)f(X)/g(X)$. In other words, we could construct a Monte Carlo estimator of $\theta = E(h(X))$ by first computing i.i.d. random vectors X_i with p.d.f. $g(X)$, then using the estimator

$$\hat{\theta} = \frac{1}{n} \sum_{i=1}^{n} \frac{h(X_i)f(X_i)}{g(X_i)}.$$

If a density $g(x)$ can be chosen so that the random variable $h(X)f(X)/g(X)$ has a small variance, then this approach is known as the importance sampling approach and can result in an efficient estimator of θ.

To see how it works, note that the ratio $f(X)/g(X)$ represents the likelihood ratio of obtaining X with respective densities f and g. If X is distributed according to g, then $f(X)$ would be small relative to $g(X)$, and therefore when X is simulated according to g, the likelihood ratio $f(X)/g(X)$ will usually be small in comparison to 1. On the other hand, it can be seen that

$$E_g\left[\frac{f(X)}{g(X)}\right] = \int \frac{f(x)}{g(x)}g(x)\,dx = \int f(x)\,dx = 1.$$

Thus, although the likelihood ratio $f(X)/g(X)$ is smaller than 1, its mean is equal to 1, suggesting that it occasionally takes large values and results in a large variance.

To make the variance of $h(X)f(X)/g(X)$ small, we arrange for a density g such that those values of X for which $f(X)/g(X)$ is large are precisely the values for which $h(X)$ is small, thus making the ratio $h(X)f(X)/g(X)$ stay small. Because importance sampling requires h to be small sometimes, it works best when estimating a small probability. Further discussions on importance sampling and likelihood method are given in Glasserman (2003).

Example 8.10   *Consider the problem* $\theta = E(U^5)$.

Suppose that we use the standard method $\hat{\theta} = \frac{1}{n}\sum_{i=1}^{n} U_i^5$, then we oversample the data near the origin and undersample the data near 1. It is easy to compute that

$$\text{Var}(\hat{\theta}) = \frac{1}{n}\{EU^{10} - (EU^5)^2\} = \frac{1}{n}(\frac{1}{11} - \frac{1}{36}) = \frac{0.0631}{n}.$$

Now, suppose we use the importance sampling, putting more weights near 1. Let $g(x) = 5x^4$ for $0 < x < 1$. Then

$$\theta_I = E_g\left(\frac{X^5 \cdot 1}{5X^4}\right) = \frac{E_g X}{5}.$$

The variance of this method is

$$\text{Var}(\hat{\theta}_I) = \frac{1}{25n}\{E_g X^2 - (E_g X)^2\}$$

$$= \frac{1}{25n}\left\{\int_0^1 x^2(5x^4)\,dx - \left(\int_0^1 x(5x^4)\,dx\right)^2\right\}$$

$$= \frac{1}{25n}\left\{5\int_0^1 x^6\,dx - (5\int_0^1 x^5\,dx)^2\right\}$$

$$= \frac{1}{25n}\left\{\frac{5}{7} - (\frac{5}{6})^2\right\}$$

$$= \frac{0.00794}{n},$$

resulting a variance reduction of 98.74%.

How do we choose g in general? This requires the notion of the so-called tilted density. Recall that the notation $M(t) = E(e^{tX})$ represents the moment-generating function (m.g.f.) of the random variable X with density f.

Definition 8.2   *A density function*

$$f_t(x) = \frac{e^{tx}f(x)}{M(t)}$$

is called a tilted density of a given f, $-\infty < t < \infty$.

Note that from this definition, a random variable with density f_t tends to be larger than the one with density f when $t > 0$, and tends to be smaller when $t < 0$.

Example 8.11   *Let f be a Bernoulli density with parameter p. Then* $f(x) = p^x(1 - p)^{1-x}$, $x = 0, 1$. *In this case, the m.g.f. is* $M(t) = E(e^{tX}) = pe^t + (1 - p)$ *so that*

$$f_t(x) = \frac{1}{M(t)} e^{tx}f(x)$$

$$= \frac{1}{M(t)} (pe^t)^x (1 - p)^{1-x}$$

$$= \left(\frac{pe^t}{pe^t + 1 - p}\right)^x \left(\frac{1 - p}{pe^t + 1 - p}\right)^{1-x}.$$

Thus, the tilted density f_t *is a Bernoulli density with parameter* $p_t = pe^t/(pe^t + 1 - p)$.

In many instances, we are interested in sums of independent random variables. In these cases, the joint density $f(x)$ of $x = (x_1, \dots, x_n)$ can be written as the product of the marginals f_i of x_i so that

$$f(x) = f_1(x_1) \cdots f_n(x_n).$$

In this situation, it is often useful to generate the X_i according to their tilted densities with a common t.

Example 8.12   *Let* $X_1, \dots, X_n$ *be independent with marginal densities* f_i. *Suppose that we are interested in estimating the quantity*

$$\theta = P(S \geq a),$$

where $S = \sum_{i=1}^n X_i$ *and* $a > \sum_{i=1}^n E(X_i)$ *is a given constant. We can apply tilted densities to estimate* θ. *Let* $I\{S \geq a\}$ *equal 1 if* $S \geq a$ *and 0 otherwise. Then*

$$\theta = E(I\{S \geq a\}),$$

where the expectation is taken with respect to the joint density. Suppose that we simulate X_i *according to the tilted density function* $f_{t,i}$, *where the value of* $t > 0$ *is to be* *specified. To construct the importance sampling estimator, note that $h(X) = I\{S \geq a\}$, $f(X) = \prod f_i(X_i)$, and $g(X) = \prod f_{t,i}(X_i)$. The importance sampling estimator would be*

$$\hat{\theta} = I\{S \geq a\} \prod_i \frac{f_i(X_i)}{f_{t,i}(X_i)}.$$

Now $f_i(X_i)/f_{t,i}(X_i) = M_i(t)e^{-tX_i}$, therefore,

$$\hat{\theta} = I\{S \geq a\} \prod_i M_i(t)e^{-tX_i}$$

$$= I\{S \geq a\}M(t)e^{-tS}, \quad (M(t) = \prod_i M_i(t)).$$

As it is assumed that $t > 0$, $S \geq a$ iff $e^{-tS} \leq e^{-ta}$ and

$$I\{S \geq a\}e^{-tS} \leq e^{-ta},$$

so that

$$\hat{\theta} \leq M(t)e^{-ta}.$$

We now find $t > 0$ such that the right-hand side of the aforementioned inequality is minimized. In that case, we obtain an estimator that lies between 0 and $\min_t M(t)e^{-ta}$. It can be shown that such t can be found by solving the equation

$$E_t(S) = a.$$

After solving for t, it can be used in the simulation. To be specific, suppose $X_1, \ldots, X_n$ are i.i.d. Bernoulli trials with $p = p_i = 0.4$. Let $n = 20$ and $a = 16$. Then

$$\hat{\theta} = I\{S \geq a\}e^{-tS} \prod_i (pe^t + 1 - p).$$

Recall from the preceding example that the tilted density $f_{t,i}$ is the p.d.f. of a Bernoulli trial with parameter $p^* = pe^t/(pe^t + 1 - p)$. It follows that

$$E_t(S) = 20p^* = \sum_{i=1}^{20} \frac{pe^t}{pe^t + 1 - p}.$$

Plugging in $n = 20, p = 0.4, a = 16$, we have

$$20\frac{0.4e^t}{0.4e^t + 0.6} = 16,$$

which leads to $e^{t^*} = 6$. Therefore, we should generate Bernoulli trials with parameter $0.4e^{t^*}/(0.4e^{t^*} + 0.6) = 0.8$ as the g and evaluate $M(t^*) = (0.4e^{t^*} + 0.6)^{20}$ and $e^{-t^*S} = (1/6)^S$. The importance sampling estimator is now

$$\hat{\theta} = I\{S \geq 16\}M(t^*)e^{-t^*S} = I\{S \geq 16\}3^{20}(1/6)^S.$$

Furthermore, we know that

$$\hat{\theta} \leq M(t^*)e^{-t^*a} = 3^{20}(1/6)^{16} = 0.001236.$$

Thus, in each iteration, the value of the importance sampling estimator lies between 0 and 0.001236.

On the other hand, we can also evaluate $\theta = P(S \geq 16)$ exactly, which equals to the probability that a Binomial random variable with parameters 20 and 0.4 be at least as big as 16. This value turns out to be 0.000317. Recall the function $h(X) = I\{S \geq 16\}$. This is a Bernoulli trial with parameter $\theta = 0.000317$. Therefore, if we simulate directly from Xs, the standard estimator $\hat{\theta}_S$ has variance

$$\text{Var}(\hat{\theta}_S) = \theta(1 - \theta) = 3.169 \times 10^{-4}.$$

As $0 \leq \hat{\theta} \leq 0.001236$, it can be shown that

$$\text{Var}(\hat{\theta}) \leq (0.001236)^2/4 = 3.819 \times 10^{-7},$$

which is much smaller than the variance of the standard estimator $\hat{\theta}_S$.

Another application of importance sampling is to estimate tail probabilities (recall at the beginning we mentioned that importance sampling works best in small probability). Suppose that we are interested in estimating $P(X > a)$, where X has p.d.f. f and a is a given constant. Let $I(X > a) = 1$ if $X > a$ and 0 otherwise. Then

$$P(X > a) = E_f(I(X > a))$$

$$= E_g\left[I(X > a)\frac{f(X)}{g(X)}\right]$$

$$= E_g\left[I(X > a)\frac{f(X)}{g(X)}|X > a\right]P_g(X > a)$$

$$+ E_g\left[I(X > a)\frac{f(X)}{g(X)}|X \leq a\right]P_g(X \leq a)$$

$$= E_g\left[\frac{f(X)}{g(X)}|X > a\right]P_g(X > a).$$

Take $g(x) = \lambda e^{-\lambda x}$, $x > 0$, an exponential density with parameter λ. Then the afore-mentioned derivation shows

$$P(X > a) = E_g[e^{\lambda X}f(X)|X > a]e^{-\lambda a}/\lambda.$$

Using the so-called "memoryless property," that is, $P(X > s + t|X > s) = P(X > t)$, of an exponential distribution, it can be easily seen that the conditional distribution of an exponential distribution conditioned on $\{X > a\}$ has the same distribution as $a + X$. Therefore,

$$P(X > a) = \frac{e^{-\lambda a}}{\lambda}E_g[e^{\lambda(X+a)}f(X + a)]$$

$$= \frac{1}{\lambda}E_g[e^{\lambda X}f(X + a)].$$

We can now estimate θ by generating $X_1, \ldots, X_n$ according to an exponential distribution with parameter λ and using

$$\hat{\theta} = \frac{1}{\lambda}\frac{1}{n}\sum_{i=1}^{n}e^{\lambda X_i}f(X_i + a).$$

Example 8.13   *Suppose that we are interested in $\theta = P(X > a)$, where X is standard normal. Then f is the normal density. Let g be an exponential density with $\lambda = a$. Then*

$$P(X > a) = \frac{1}{a}E_g[e^{aX}f(X + a)]$$

$$= \frac{1}{a\sqrt{2\pi}}E_g[e^{aX-(X+a)^2/2}].$$

We can therefore estimate θ by generating X, an exponential distribution with rate a, and then using

$$\hat{\theta} = \frac{e^{-a^2/2}}{a\sqrt{2\pi}}\frac{1}{n}\sum_{i=1}^{n}e^{-X_i^2/2}$$

to estimate θ. To compute the variance of $\hat{\theta}$, we need to compute quantities $E_g[e^{-X^2/2}]$ and $E_g[e^{-X^2}]$. These can be computed numerically and can be shown to be

$$E_g[e^{-X^2/2}] = ae^{a^2/2}\sqrt{2\pi}(1 - \Phi(a)), \quad E_g[e^{-X^2}] = ae^{a^2/4}\sqrt{\pi}(1 - \Phi(a/\sqrt{2})).$$

For example, if $a = 3$ and $n = 1$, then $\mathrm{Var}(e^{-X^2/2}) = 0.0201$ and $\mathrm{Var}(\hat{\theta}) = (\frac{e^{-4.5}}{3\sqrt{2\pi}})^2 \times$ $0.0201 \sim 4.38 \times 10^{-8}$. *On the other hand, a standard estimator has variance $\theta(1 - \theta) = 0.00134$.*

Consider simulating a vanilla European call option price again, using the importance sampling technique. Suppose that we evaluate the value of a deep out-of-money $(S_0 \ll K)$ European call option with a short maturity T. Many sampling paths result

$S_T \leq K$ and give zero-values. Thus, these samples are wasted. One possible way to deal with this problem is to increase the values of Z_is by sampling them from a distribution with large mean and large variance. Sample $\tilde{Z}_i$ from $N(\frac{m}{\sigma\sqrt{T}}, s^2)$ so that

$$\sigma\sqrt{T}\tilde{Z}_i \sim N(m, \sigma^2 T s^2).$$

Note that $\tilde{Z}_i$ can be written as

$$\tilde{Z}_i = \frac{m}{\sigma\sqrt{T}} + s Z_i, \quad Z_i \sim N(0, 1).$$

The importance sampling estimator is then given by

$$C_I = e^{-rT}\frac{1}{N}\sum_{i=1}^{N} \max\{S_0 e^{(r-\sigma^2/2)T+\sigma\sqrt{T}\tilde{Z}_i} - K, 0\}R(\tilde{Z}_i),$$

where

$$R(\tilde{Z}_i) = \frac{\frac{1}{\sqrt{2\pi}}\exp(-\tilde{Z}_i^2/2)}{\frac{1}{\sqrt{2\pi}s}\exp(-\frac{1}{2s^2}(\tilde{Z}_i - \frac{m}{\sigma\sqrt{T}})^2)} = s\exp(\frac{Z_i^2}{2} - \frac{\tilde{Z}_i^2}{2}).$$

Thus, C_I can be expressed as

$$C_I = se^{-rT}\frac{1}{N}\sum_{i=1}^{N}\max\{S_0 e^{(r-\sigma^2/2)T+m+s\sigma\sqrt{T}\tilde{Z}_i} - K, 0\}\exp\left(\frac{Z_i^2}{2} - \frac{(\frac{m}{\sigma\sqrt{T}}+sZ_i)^2}{2}\right).$$

Example 8.14 *Let $S_0 = 100$, $K = 140$, $r = 0.05$, $\sigma = 0.3$, and $T = 1$. We simulate the value of this deep out-of-money European call option, using the importance sampling technique and compare it with the result of standard method. See the online material for the VBA codes.*

For $N = 10,000$, we have $\overline{C}_I = 3.1202$ with standard error 0.0264 using importance sampling while getting $\overline{C} = 3.0166$ with standard error 0.1090 using standard method. The result shows that the importance sampling technique gives a more precise estimate of the price of the option, which has a theoretical Black–Scholes price 3.1187.

8.6 EXERCISES

1. Let $U \sim U(0, 1)$ and let a and b be two given constants with $a < b$. Show that $Y = a + (b - a)U$ is distributed as a $U(a, b)$ random variable.

2. Let $\hat{b}$ be the least squares estimate of b in the simple linear regression model $X = a + bY + e$, $e \sim (0, \sigma^2)$ i.i.d.. Show that

$$\mathrm{Var}(X - \hat{b}Y) = \mathrm{Var}(X) - \hat{b}^2\mathrm{Var}(Y).$$

3. Suppose that you want to estimate $\theta = \int_0^1 e^{x^2}\,dx$. Show that generating a random number U and then using the antithetic estimator $(e^{U^2}(1 + e^{1-2U}))/2$ is better than generating two random numbers U_1 and U_2 and using the standard estimator $(e^{U_1^2} + e^{U_2^2})/2$.

4. Consider estimating $\theta = \int_0^1 4x^3\,dx$.
   (a) Using standard simulation technique, estimate θ.
   (b) Using antithetic variable technique, construct an improved estimate of θ.
   (c) Using stratification, construct another estimate of θ.
   (d) Construct a control variate estimate of θ.
   (e) Compare the performance of these different estimates.
   (f) Can you combine the aforementioned methods to improve the result?

5. Consider $\theta = \int_2^\infty (x - 2)e^{-x}\,dx$.
   (a) It is known that $\theta = E[f(X)]$ where $X \sim \mathrm{Exp}(1)$. What is $f(X)$?
   (b) Provide an algorithm to sample X from the interval $[2, \infty)$.
   (c) Provide an algorithm to stratify X in the interval $[2, \infty)$ with equal probability $1/4$ for each stratified interval.
   (d) Provide a Monte Carlo algorithm using $(X - 2)$ as the control variable.

6. Redo Examples 8.6, 8.7, and 8.9 using $S_0 = K = 100$, $r = 0.05$, $\sigma = 0.1$, and $T = 1$. Calculate the theoretical Black–Scholes price as well.

7. Verify Equations 8.4 and 8.5.

8. Consider a *truncated payoff* vanilla call option with maturity T and strike price K. The payoff function is given by

$$h(S_T) = \begin{cases} S_T - K & \text{if } K \le S_T \le S_b, \\ 0 & \text{otherwise .} \end{cases}$$

The given constant S_b acts as a barrier, canceling the option whenever $S_T > S_b$. Assuming that the stock price follows a geometric Brownian motion with $v = r - \sigma^2/2$, where the risk-free rate r and the volatility σ are known. Using the idea of antithetic variables, write a variance reduction algorithm to estimate the payoff function.

The solutions and/or additional exercises are available online at http://www.sta.cuhk.edu.hk/Book/SRMS/.

9

PATH DEPENDENT OPTIONS

9.1 INTRODUCTION

Contingent claims other than standard call and put options are known as exotic options. The most common type of exotic options is path dependent options. As indicated by the name, the payoff of a path dependent option depends on the entire path of the underlying asset prices, not just the terminal asset price alone. According to this definition, American options are path dependent options because the option holder has to determine whether the options are worth to exercise at each time point. The path-dependent feature of an option usually complicates the analytical tractability of valuation. Simulation would be the most useful alternative.

Owing to the need to value exotic options, this chapter studies simulation techniques for European and American style path dependent options. Some of the options considered in this chapter have no analytical solutions.

9.2 BARRIER OPTION

Barrier options have become increasingly popular nowadays. A barrier option is very much similar to a "vanilla" option, which becomes alive when the barrier is crossed. Let K be the strike price, T be the time to maturity, and V be the value of the barrier. A **down-and-in** barrier option becomes **alive** only if the stock price (usually counting only closing prices) goes below V before T. A **down-and-out** barrier option is

Simulation Techniques in Financial Risk Management, Second Edition. Ngai Hang Chan and Hoi Ying Wong.
© 2015 John Wiley & Sons, Inc. Published 2015 by John Wiley & Sons, Inc.

killed if the stock price goes below V before T. A down-and-in barrier call option is a cheaper tool to hedge against the upside risk. From the definition, it can be easily seen that holding both a down-and-in and down-and-out options with the same strike price K and maturity T is the same as holding a "vanilla" option. Let C_{di} and C_{do} be the option value of the down-and-in call and the down-and-out call, respectively. Then

$$C_{di} + C_{do} = C,$$

where C is the vanilla call price. Let

$$S_{min} = \min_{0 < t \leq T} S(t) \quad \text{and} \quad I\{S_{min} < V\} = \begin{cases} 1 & S_{min} < V, \\ 0 & S_{min} \geq V, \end{cases}$$

be the realized minimum asset price and the indicator of the down-and-in option, respectively. Then, the value of the option can be written as

$$C_{di} = e^{-rT} \hat{E}\{I\{S_{min} < V\}(S(T) - K)^+\},$$

where $\hat{E}$ denotes the risk-neutral expectation. The other types of barrier options can be evaluated analogously.

To simulate the value of a down-and-in call option, the algorithm goes as follows:

1. Generate the daily stock price $S(t_1), S(t_2), \ldots, S(t_n = T)$. If $\min_i S(t_i) < V$, then set

$$C = e^{-rT} \max(S(T) - K, 0),$$

   else set $C = 0$.

2. Repeat Step 1 N times to obtain $C_1, \ldots, C_N$. The value of the down-and-in call option is given by

$$\overline{C} = \frac{1}{N} \sum_{i=1}^{N} C_i,$$

   and the standard error of the estimator is given by

$$\sqrt{\frac{1}{N(N-1)} \sum_{i=1}^{N} (C_i - \overline{C})^2}.$$

Example 9.1 *Let $S_0 = 10$, $r = 0.23$, $\sigma = 0.4$, and $dt = 1/250$. Compute the value of a down-and-in call option with strike price $K = 12$, maturity $T = 1$, and barrier $V = 9$. Please see the online material for the VBA codes.*

For $N = 10,000$, we get $\overline{C} = 1.0273$, and the standard error of $\overline{C}$ is 0.02048. The 95% confidence interval for C is $[0.9872, 1.0675]$.

9.3   LOOKBACK OPTION

The payoffs of lookback options depend on the maximum or the minimum stock price during the life of the option. Denote the maximum (minimum) of the stock price over the time period $[0, T]$ by $S_{max}(T)$ ($S_{min}(T)$). Four popular lookback options are as follows:

1. Floating strike lookback call (c_{fl}): payoff $= S_T - S_{min}(T)$;
2. Floating strike lookback put (p_{fl}): payoff $= S_{max}(T) - S_T$;
3. Fixed strike lookback call (c_{fix}): payoff $= \max(S_{max}(T) - K, 0)$;
4. Fixed strike lookback put (p_{fix}): payoff $= \max(K - S_{min}(T), 0)$.

There are lookback put-call parities connecting the floating strike lookback call (put) to the fixed strike lookback put (call). Specifically, four put-call parities of lookback options are as follows:

1. $c_{fl}\left(t, S, S_{min}(t)\right) = S - e^{-r(T-t)}S_{min}(t) + p_{fix}\left(t, S, S_{min}(t); K = S_{min}(t)\right)$;
2. $p_{fl}\left(t, S, S_{max}(t)\right) = e^{-r(T-t)}S_{max}(t) - S + c_{fix}\left(t, S, S_{max}(t); K = S_{max}(t)\right)$;
3. $c_{fix}\left(t, S, S_{max}(t); K\right) = S - e^{-r(T-t)}K + p_{fl}\left(t, S, \max(S_{max}(t), K)\right)$;
4. $p_{fix}\left(t, S, S_{min}(t); K\right) = e^{-r(T-t)}K - S + c_{fl}\left(t, S, \min(S_{min}(t), K)\right)$.

These four put-call parities are model independent, meaning that they are applicable to any asset dynamics. For a proof, we refer to the article of Wong and Kwok (2003).

Pricing lookback options with simulation is very similar to that of the barrier option. Consider the floating strike lookback call option. The VBA code of Example 9.1 can be modified to obtain the lookback option price. We just compute

$$\frac{e^{-rT}}{N} \sum_{i=1}^{N} \left[S_i(T) - \min_{j} S_i(t_j) \right].$$

Other lookback options are valued in the same manner.

It is interesting to notice that simulating fixed strike lookback options requires less storage than simulating the floating ones. The reason is that payoffs of fixed strike lookback options do not depend on the terminal asset price, S_T. Therefore, after generating a sample path, only the maximum or minimum price of the path is required. With this observation and the lookback put-call parities, a storage-saving approach to simulating floating strike lookback options can be developed. For valuing a floating strike lookback call, a fixed strike lookback put with strike price equaling to the realized minimum asset value is simulated. Then, the floating strike call price is extracted from the first put-call parity.

9.4 ASIAN OPTION

Asian options payoffs depend on the average of the underlying asset prices during the option life. Asian options are popular in the financial industry because they cost less than their vanilla counterparts and are less sensitive to the change in underlying asset prices. The common forms of averaging in option contracts can be either geometric average or arithmetic average of the underlying variables. Denote the geometric average and arithmetic average of the underlying asset in the period $[0, T]$ by G_T and A_T, respectively. Then,

$$G_T = \lim_{n \to \infty} \left[\prod_{i=1}^{n} S(t_i) \right]^{\frac{1}{n}} = \exp \left[\frac{1}{T} \int_0^T \log S(t)\, dt \right], \tag{9.1}$$

$$A_T = \lim_{m \to \infty} \frac{1}{n} \sum_{i=1}^{n} S(t_i) = \frac{1}{T} \int_0^T S(t)\, dt.$$

For geometric Asian options, analytical pricing formulas are available in the literature; see for example Wong and Cheung (2004). However, almost all Asian options are traded with arithmetic average. For instance, two frequently traded Asian options are as follows:

1. Floating strike Asian call. Payoff $= \max(S_T - A_T, 0)$;
2. Fixed strike Asian call. Payoff $= \max(A_T - K, 0)$.

In practice, the geometric Asian option prices are used as a control variate in simulating their arithmetic counterparts.

Let us illustrate the procedure by considering a fixed strike Asian call. The geometric version of the option has the payoff $\max(G_T - K, 0)$. Denote X_T by $\log G_T$, that is,

$$X_T = \frac{1}{T} \int_0^T \log S(\tau)\, d\tau.$$

By Itô's lemma,

$$\log S_\tau = \log S_t + v(\tau - t) + \sigma(W(\tau) - W(t)) \quad \text{(Recall: } v = r - \sigma^2/2\text{)},$$

which implies

$$X_T = X_t \frac{t}{T} + \frac{1}{T} \int_t^T \log S(\tau)\, d\tau$$
$$= X_t \frac{t}{T} + \frac{T-t}{T} \log S_t + v \frac{(T-t)^2}{2T} + \frac{\sigma}{T} \left[\int_t^T W(\tau)\, d\tau - (T-t)W_t \right]$$

$$= X_t \frac{t}{T} + \frac{T-t}{T} \log S_t + v \frac{(T-t)^2}{2T} + \frac{\sigma}{T} \left[T(W_T - W_t) - \int_t^T \tau \, dW_\tau \right]$$

$$= X_t \frac{t}{T} + \frac{T-t}{T} \log S_t + v \frac{(T-t)^2}{2T} + \frac{\sigma}{T} \int_t^T (T - \tau) \, dW_\tau,$$

where the second last line uses the integration by parts formula; see Example (4.2). By Itô's identities, see Exercise 1(d) in Chapter 4, we have

$$\mathrm{E} \int_t^T (T - \tau) \, dW(\tau) = 0 \quad \text{and}$$

$$\mathrm{Var} \left[\int_t^T (T - \tau) \, dW(\tau) \right] = \int_t^T (T - \tau)^2 \, d\tau = \frac{(T-t)^3}{3}.$$

Therefore,

$$X_T \sim N \left(\frac{t}{T} X_t + \frac{T-t}{T} \log S_t + v \frac{(T-t)^2}{2T}, \sigma^2 \frac{(T-t)^3}{3T^2} \right). \tag{9.2}$$

Risk-neutral valuation asserts that

$$C_G^{fix}(t, S, G_t) = e^{-r(T-t)} \hat{\mathrm{E}} \left[\max(e^{X_T} - K, 0) \right].$$

Applying Lemma 5.1, we obtain the closed form solution as

$$C_G^{fix}(t, S, G_t) = S \left(\frac{G_t}{S} \right)^{\frac{t}{T}} e^{R(t,T)} \Phi(\hat{d}_1) - K e^{-r(T-t)} \Phi(\hat{d}_2), \tag{9.3}$$

where

$$\hat{d}_1 = \frac{T \log \frac{S}{K} + t \log \frac{G_t}{S} + (r - \frac{\sigma^2}{2}) \frac{(T-t)^2}{2} + \sigma^2 \frac{(T-t)^3}{3T}}{\sqrt{\sigma^2 \frac{(T-t)^3}{3}}}, \tag{9.4}$$

$$\hat{d}_2 = \hat{d}_1 - \sqrt{\sigma^2 \frac{(T-t)^3}{3T^2}},$$

$$R(t; T) = \left(r - \frac{\sigma^2}{2} \right) \frac{(T-t)^2}{2T} + \sigma^2 \frac{(T-t)^3}{6T^2} - r(T-t).$$

With the analytical solution of the geometric Asian call (GAC), we simulate the arithmetic Asian price via control variate. The algorithm is presented as follows.

Step 1: Generate daily stock prices $S(t_1), S(t_2), \dots, S(t_n)$.
Step 2: Set

$$G_j = \left[\prod_{i=1}^n S(t_i) \right]^{\frac{1}{n}}, \quad C_G^j = e^{-r(T-t)} \max(G_j - K, 0),$$

$$A_j = \frac{1}{n} \sum_{i=1}^{n} S(t_i), \quad C_A^j = e^{-r(T-t)} \max(A_j - K, 0).$$

Step 3: Repeat Steps 1 and 2 N times.

Step 4: Compute the regression coefficients a and b by fitting

$$C_A^j = a + b\, C_G^j, \quad j = 1, 2, \ldots, N.$$

Step 5: $C_A^{fix} = a + b\, C_G^{fix}(t, S, G_t)$ with formula (9.3) applied.

Example 9.2 *Consider the parameter values:* $S_t = 10, r = 0.03, \sigma = 0.4, t = 0.2, T = 1,$ *and the realized arithmetic average* $A_t = 10.5.$ *Simulate the arithmetic Asian call option with a fixed strike price of $12. Please see the online material for the VBA implementation.*

This simulation gives the arithmetic Asian call (AAC) price to be 0.1698. The analytical price for the GAC is computed as 0.1318. The AAC is a bit more expensive than the GAC because the arithmetic mean always dominates the geometric mean. The computational time is about 10 s.

9.5   AMERICAN OPTION

American options allow the holder to exercise before maturity. This early exercise feature exists in major financial markets. The valuation and optimal exercise of American options is one of the most challenging problems in derivatives finance, especially when more than one factor is involved in the option contract.

Although simulation techniques can be used to generate future scenarios, the forward looking feature of simulation complicates the valuation of American option, where optimal exercising policy have to be constructed via backward reduction. When an American put option is valued with binomial tree, one has to determine if it is optimal to exercise the option at each node in a backward manner. A practical approach to valuing American options with simulation is proposed by Longstaff and Schwartz (2001). This section presents the idea of American option pricing using this approach.

9.5.1   Simulation: Least Squares Approach

The best way to illustrate the least squares approach of Longstaff and Schwartz (2001) is by means of a concrete example. In the following numerical example, we introduce the algorithm in detail first and explain the concepts later.

Example 9.3 *Let* $S(0) = 10, r = 0.03, \sigma = 0.4.$ *Compute the value of an American put option with strike price* $K = 12$ *and maturity* $T = 1.$ *For simplicity, assume that the option can be exercised at* $t = 1/3, 2/3,$ *and* $1.$

TABLE 9.1   Sample Paths

Path	$t = 1/3$	$t = 2/3$	$t = 1$	Y_3 $= \max(K - S(1), 0)$
1	8.3826	9.9528	6.7581	5.2419
2	11.9899	13.8988	14.5060	0
3	13.1381	17.4061	13.4123	0
4	6.8064	7.8115	10.6520	1.3480
5	7.0508	9.1293	7.4551	4.5449
6	11.2214	8.3600	9.2896	2.7104
7	8.9672	8.7787	9.0822	2.9178
8	11.5336	10.9398	8.6958	3.3042

TABLE 9.2   Regression at $t = 2/3$

Path	$Y_3 e^{-r\Delta t}$	$S(2/3)$	Exercise in-the-Money?
1	5.1898	9.9528	Yes
2	—	13.8988	No
3	—	17.4061	No
4	1.3346	7.8115	Yes
5	4.4997	9.1293	Yes
6	2.6834	8.3600	Yes
7	2.8888	8.7787	Yes
8	3.2714	10.9398	Yes

We use the formula $S(t + \Delta t) = S(t) \exp[(r - \sigma^2/2)\Delta t + \sigma \Delta W_t]$ to generate asset prices at exercise time points: $t = 1/3, 2/3$, and 1. Table 9.1 gives eight sample paths. Terminal payoffs corresponding to each path, Y_3, are given by the last column of the table. Discounting the sample mean of the terminal payoffs estimates the European put price to be \$2.4343. This is a lower bound for the American put option.

At time $t = 2/3$, the option holder must decide whether to exercise the option immediately or to continue the option when the option is in-the-money. To make the decision, the holder should compare the cash flows of immediate exercise with the expected payoff of continuation given the asset price at time 2/3. Therefore, it is essential to estimate the conditional expected payoff. To do this, we collect the response variable $Y_3 e^{-r\Delta t}$ and the explanatory variable $S(2/3)$ for in-the-money paths in Table 9.2, where $\Delta t = 1/3$. We model the expected payoff from continuation at time $t = 2/3$ as a quadratic polynomials, $f_2(S_t)$, of asset values at time $t = 2/3$. Coefficients of the polynomials are estimated from the data in Table 9.2 by the least squares method. Therefore, we estimate $\hat{a}_0, \hat{a}_1$ and $\hat{a}_2$ from the regression line:

$$Y_3 e^{-r\Delta t} = \hat{a}_0 + \hat{a}_1[S(2/3)] + \hat{a}_2[S(2/3)]^2 + \epsilon.$$

The resulting formula is

$$E[Y_3 e^{-r\Delta t}|S(2/3)] = -82.5347 + 17.7788[S(2/3)] - 0.9063[S(2/3)]^2 := f_2(S).$$

With this conditional expectation function, $f_2(S)$, we are able to compare the value of immediate exercise, $K - S(2/3)$, and compute payoffs, Y_2, for each path at $t = 2/3$. The value of Y_2 is obtained by the formula,

$$Y_2 = \begin{cases} K - S(2/3), & \text{if } K - S(2/3) \geq f_2(S(2/3)), \\ e^{-\Delta t} Y_3, & \text{otherwise.} \end{cases}$$

This formula asserts that the payoff at time $t = 2/3$ is $K - S$ if exercising the option is worth more than the expected payoff from holding it; otherwise, the payoff at time 2/3 becomes the discounted cash flow in the next exercise time. The last column of Table 9.3 gives the expected payoffs, Y_2, for each sample path.

Next, we repeat the procedure for $t = 1/3$. In Table 9.4, all sample paths are in-the-money except path 3. Then, the least squares estimation corresponding to in-the-money paths gives

$$E[Y_2 e^{-r\Delta t}|S(1/3)] = -8.9488 + 3.3104S(1/3) - 0.2036[S(1/3)]^2 := f_1(S).$$

This regression function determines the exercising policy at $t = 1/3$.

TABLE 9.3   Optimal Decision at $t = 2/3$

Path	Exercise $K - S(2/3)$	Continuation $f_2(S(2/3))$	$e^{-r\Delta t}Y_3$	Y_2
1	2.0472	4.6380	5.1898	5.1898
2	—	—	0	0
3	—	—	0	0
4	4.1885	1.0428	1.3346	4.1885
5	2.8707	4.2388	4.4997	4.4997
6	3.6400	2.7554	2.6834	3.6400
7	3.2213	3.6959	2.8888	2.8888
8	1.0602	3.4968	3.2714	3.2714

TABLE 9.4   Regression at $t = 1/3$

Path	$Y_2 e^{-r\Delta t}$	$S(1/3)$	Exercise in-the-Money?
1	5.1381	8.3826	Yes
2	0	11.9899	Yes
3	0	13.1381	No
4	4.1468	6.8064	Yes
5	4.4549	7.0508	Yes
6	3.6038	11.2214	Yes
7	2.8600	8.9672	Yes
8	3.2388	11.5336	Yes

TABLE 9.5 Optimal Decision at $t = 1/3$

Path	Exercise $K - S(1/3)$	Continuation $f_1(S(1/3))$	$e^{-r\Delta t}Y_2$	Y_1
1	3.6174	4.4921	5.1381	5.1381
2	0.0101	1.4689	0	0
3	—	—	0	0
4	5.1936	4.1494	4.1468	5.1936
5	4.9492	4.2688	4.4549	4.9492
6	0.7786	2.5572	3.6038	3.6038
7	3.0328	4.3620	2.8600	2.8600
8	0.4664	2.1440	3.2388	3.2388

Once again, the Y_1 in Table 9.5 is computed according to the optimal decision by the rule,

$$Y_1 = \begin{cases} K - S(1/3), & \text{if } K - S(1/3) \geq f_1(S(1/3)), \\ e^{-\Delta t}Y_2, & \text{otherwise.} \end{cases}$$

Finally, the current price of the American option is estimated by the average of $e^{-r\Delta t}Y_1$, that is, \$3.0919, which is higher than the European option price \$2.4343.

9.5.2 Analyzing the Least Squares Approach

Consider an American put option with exercise rights at $t_1 < \cdots < t_n = T$. To simplify matters, we assume $t_{j+1} - t_j = \Delta t$ for $j = 1, 2, \ldots, n - 1$. Given a sample path of the underlying asset price, $\{S(t_1), S(t_2), \ldots, S(t_n)\}$, we study possible payoffs received by the option holder at each of the exercise time points. Clearly, if the option is not exercised prematurely, then the holder receives the terminal payoff, denoted as $Y_n = \max(K - S(t_n), 0)$. At time $t = t_{n-1}$, the corresponding payoff, Y_{n-1}, depends on the holder's decision of exercising the option. Therefore,

$$Y_{n-1} = \begin{cases} K - S(t_{n-1}), & \text{exercise,} \\ e^{-r\Delta t}Y_n, & \text{continue.} \end{cases}$$

This formula indicates that the option holder receives $K - S(t_{n-1})$ if the optimal decision is to exercise the option. Otherwise, the holder will receive a cash flow of Y_n at the next time step. The present value of this cash flow is obtained through multiplying a discounted factor $e^{-r\Delta t}$. Inductively, the payoff Y_j at time t_j can be described as

$$Y_j = \begin{cases} K - S(t_j), & \text{exercise,} \\ e^{-r\Delta t}Y_{j+1}, & \text{continue.} \end{cases} \tag{9.5}$$

This iterative process stops until Y_1 is obtained. As the option holder has no exercise right in the time period $[0, t_1)$, the American put option can be viewed as a European option that expires at t_1 with payoff Y_1. Risk-neutral valuation allows us to value the American put, $P_A(0, S)$, as

$$P_A(0, S) = \hat{E}\left[e^{-rt_1}Y_1|S_0 = S\right].$$

Therefore, a typical simulation algorithm generates N sample paths; each follows the algorithm to obtain $\{Y_1^{(1)}, \ldots, Y_1^{(N)}\}$. The American put is estimated by

$$P_A(0, S) = \frac{1}{N}\sum_{i=1}^{N}e^{-rt_1}Y_1^{(i)}. \tag{9.6}$$

The aforementioned simulation is incomplete, however. To simulate the American put, the payoff, Y_1, at time t_1 should be obtained via simulation. This requires the simulation algorithm to detect optimal exercise at each time point successively. In other words, we have to clarify the condition of exercising the option in Equation 9.5. It is crucial that the optimal decision should not be made by simply comparing the values of $K - S(t_j)$ and $e^{-r\Delta t}Y_{j+1}$ in Equation 9.5. The reason is that the decision at time t_j should be based on the information up to t_j. However, the value Y_{j+1} depends on the asset value at t_{j+1}. The correct approach is to compare the immediate exercise cash flow $K - S(t_j)$ with the expectation on the discounted cash flow conditional on the asset price $S(t_j)$. This leads (Eq. 9.5) to

$$Y_j = \begin{cases} K - S(t_j), & \text{if } K - S(t_j) \geq f_j(S(t_j)), \\ e^{-r\Delta t}Y_{j+1}, & \text{if } K - S(t_j) < f_j(S(t_j)), \end{cases} \tag{9.7}$$

where $f_j(S(t_j))$ is the conditional expectation function at t_j, that is,

$$f_j(S(t_j)) = \hat{E}\left[e^{-r\Delta t}Y_{j+1}|S(t_j)\right]. \tag{9.8}$$

The key to the Longstaff and Schwartz (2001) approach is the use of least squares to estimate the function, $f_j(S)$. Under certain technical conditions, it can be shown that the function $f_j(S(t_j))$ can be approximated by a polynomial of $S(t_j)$. In other words,

$$f_j(S(t_j)) = \sum_{k=0}^{\infty}a_k[S(t_j)]^k,$$

where $\{a_k\}$ converges to zero rapidly. Therefore, one way to approximate $f_j(S)$ is by truncating the polynomial of infinite order to a finite order polynomial. Coefficients of the finite order polynomials are estimated through the least squares method.

In Example 9.3, we use a polynomials of degree 2 to approximate $f_j(S)$. The simulation starts by generating N asset price paths, $\{S_i(t_1), \ldots, S_i(t_n)\}$ for $i = 1, 2, \ldots, N$. When $t = t_n$, it is clear that $Y_n^{(i)} = \max[K - S_i(t_n), 0]$ for the path i. We go one step back to the time point $t = t_{n-1}$, where N possible asset prices have been generated.

Then, the coefficients a_0, a_1, and a_2 are obtained by taking least squares estimation to the regression line:

$$f_{n-1}(S) = \widehat{E}\left[e^{-r\Delta t}Y_n|S\right] = a_0 + a_1[S(t_{n-1})] + a_2[S(t_{n-1})]^2. \qquad (9.9)$$

The estimation is based on the sample $\{(S_i(t_{n-1}), Y_n^{(i)})|K > S_i(t_{n-1}), i = 1, \ldots, N\}$, that is, in-the-money paths. Then, payoffs at t_{n-1} are calculated via the rule in Equation 9.7. Having a sample of payoffs $\{Y_{n-1}^{(i)}|i = 1, 2, \ldots, N\}$ at t_{n-1}, we go one step back to the time point t_{n-2} and repeat the process. Eventually, we obtain N possible payoffs, $\{Y_1^i|i = 1, 2, \ldots, N\}$, at t_1. Monte Carlo simulation estimates the current option price by the average in Equation 9.6.

Remarks

1. In the regression Equation 9.9, only in-the-money paths are used in the least squares estimation, as these paths are sensitive to immediate exercise. Remember that the option holder will exercise the option only when it is in-the-money.

2. An obvious way to improve the accuracy is to increase the number of terms in Equation 9.9. However, one has to strike a balance between increasing the number of terms and the quality of estimates. Numerical experiments show that polynomials of degree 3 are a reasonable choice.

3. Instead of using ordinary monomials as basis functions in Equation 9.9, one may consider other basis functions, such as Hermite, Laguerre, Legendre, Chebyshev, Gegenbauer, and Jacobi polynomials. Numerical tests of Moreno and Navas (2003) show that the least squares approach is quite robust to the choice of basis functions. For more complex derivatives, this choice can slightly affect option prices.

4. The recent analysis of Stentoft (2004) indicates that a modified specification using ordinary monomials is preferred over the specification based on Laguerre polynomials used in Longstaff and Schwartz (2001). Furthermore, the least squares method is computationally more efficient than other numerical methods, such as finite difference, especially when high dimensional problems are concerned.

5. The article by Longstaff and Schwartz (2001) points out that the R^2 values of the regressions are often low. This means that the volatility of unexpected cash flows is large relative to the expected cash flows. However, because the least squares simulation is based on conditional first moments rather than higher moments, the R^2's of the regression should have little impact on estimated American option price.

6. If the user is really concerned about the R^2, it may be more efficient to use other techniques such as weight least squares and generalized method of moments (GMM) in estimating the conditional expectation function.

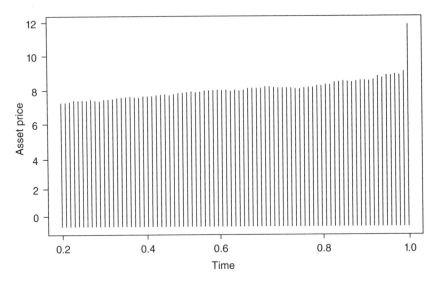

Figure 9.1   The exercising region of the American put option.

Example 9.4   *Using the parameters in the preceding example, simulate the American put price with continuous exercise rights and hence determine the optimal exercise policy. The simulation is based on 10,000 sample paths with $\Delta t = 1/100$. The online materials provide the VBA codes for the simulation.*

By using quadratic conditional expectation functions, our simulation estimates the American put price as 2.739 within 15 s, which is consistent with the binomial model of Hull (2006). For the early exercise policy, we collect the maximum asset value that belongs to the exercising region at each time. For $t \geq 0.2$, Figure 9.1 plots the exercise policy against time. The option is optimal to exercise if the stock price falls into the shaded region. It is seen that the early exercise boundary looks similar to an increasing function of calendar time and hence a decreasing function of option maturity. For $t < 0.2$, our simulation has no path in the exercising region so that we are unable to graph the exercising boundary.

9.5.3   American Style Path Dependent Options

The examples considered so far are relevant to pricing American put options; the least squares approach is applicable to any early exercisable contingent claims. Denote the terminal payoff function of a path dependent option by $F(S_T, \xi_T)$ where ξ is an exogenous variable. For instance, $\xi_T = S_{min}(T)$ for a barrier option or a lookback option and $\xi_T = A_T$ for an Asian option. The American style path dependent option with payoff $F(S_T, \xi_T)$ can be simulated as follows.

Step 1: Generate asset price paths $\{S_i(t_1), S_i(t_2), \ldots, S_i(t_n)\}$ for $i = 1, 2, \ldots, N$. Set $j = n - 1$ and $Y_n^i = F(S_i(t_n), \xi_i(t_n))$.

Step 2: Use least squares to estimate coefficients of a polynomials of degree m, $\mathcal{P}_m(S_i(t_j), \xi_i(t_j))$ from:

$$e^{-r\Delta t} Y^i_{j+1} = \mathcal{P}_m(S_i(t_j), \xi_i(t_j)),$$

for in-the-money paths.

Step 3: If $F(S_i(t_j), \xi_i(t_j)) \geq \mathcal{P}_m(S_i(t_j), \xi_i(t_j))$, then set $Y^i_j = F(S_i(t_j), \xi_i(t_j))$; otherwise, set $Y^i_j = e^{-r\Delta t} Y^i_{j+1}$.

Step 4: If $j > 1$, then set $j = j - 1$ and go to Step 2.

Step 5: The American option price $= \frac{1}{N} \sum_{i=1}^{N} e^{-r\Delta t} Y^i_1$.

Example 9.5 *Suppose that $S_0 = 10, r = 0.03, \sigma = 0.4$, and $T = 7/12$ (7 months). Simulate the American style floating strike arithmetic Asian put option and plot the optimal exercise regions for $t = 0.2, 0.4, 0.6$, and 0.8. The simulation is based on 10,000 sample paths with $\Delta t = 1/100$.*

We approximate the conditional expectation function, $f_j(S, A)$, by a two-variable quadratic polynomials, that is,

$$f_j(S, A) = a_{00} + a_{10}S + a_{20}S^2 + a_{11}SA + a_{01}A + a_{02}A^2.$$

Our simulation estimates the option price to be 9.783. This number is consistent with the one obtained by the finite difference method (FDM) in Hansen and Jorgensen (2000). The CPU (central processing unit) time is about 17 s for the computation. Figure 9.2 plots the exercise boundaries at time 0.2, 0.4, 0.6, and 0.8. The boundaries are the interfaces between shaded and nonshaded regions. The shaded regions are those of the continuation regions. For $t = 0.2$, there are less points falling into the exercising region. Thus, the simulation is only able to graph the exercise boundary for underlying asset prices in the range of 7–11 at $t = 0.2$.

9.6 GREEK LETTERS

As pointed out in Chapter 7, hedging is sometimes more important than pricing in risk management. Option hedging requires risk managers to compute option Greeks, such as delta, gamma, vega, and theta. We refer interested readers to Hull (2006) for the application of Greeks in hedging and Joshi (2003) for discrete tree approximation. The Greek letters are actually representing partial differentiations of the option pricing formula with respect to different parameters. Because most options, especially path dependent options, do not have closed form pricing formulas, Greeks should be obtained by means of simulation. For single asset path-independent options, the simulation can be constructed via Theorem 7.3. However, it is inapplicable for path dependent options. Thus, we introduce an alternative practical approach to simulating Greeks.

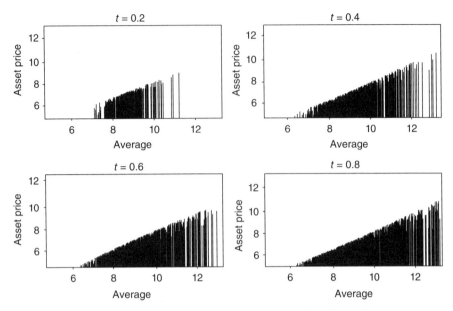

Figure 9.2   Exercise regions of the American style Asian option.

Let V denote the pricing formula of an option. The option Greeks are defined as follows.

$$\text{Delta} = \frac{\partial V}{\partial S};$$

$$\text{Gamma} = \frac{\partial^2 V}{\partial S^2};$$

$$\text{Vega} = \frac{\partial V}{\partial \sigma};$$

$$\text{Theta} = \frac{\partial V}{\partial t};$$

$$\text{Rho} = \frac{\partial V}{\partial r},$$

where S is the underlying asset price, σ is the volatility, t is the time variable, and r is the spot interest rate. Hence, the Greeks can be obtained by standard differentiation techniques or approximated by the numerical FDM if the option pricing formula is available. The FDM computes numerical differentiation by approximating the first principle in differentiation. For instance, suppose that we are interested in the Delta of an option. Then, the FDM approximates the value by

$$\text{Delta} \simeq \frac{V(S+h) - V(S)}{h}, \tag{9.10}$$

where h is an arbitrarily chosen small number and other parameters are fixed.

The approach introduced here combines simulation with FDM together. Suppose that we need the Delta of an option. Then we proceed as follows. Firstly, the option price is simulated as usual with the current realized asset price S. Secondly, we re-simulate the option price again with a "perturbed" asset price $S + h$. Finally, the Delta is approximated by Equation 9.10. However, the stability of this approach would be of great concerns because there are two sources of errors: simulation error and FDM error. The most critical one is the simulation error, which makes the numerator of Equation 9.10 nonzero even when h tends to zero. To circumvent this difficulty, it is very common for market practitioners to use the same set of random numbers in the first and the second steps. We illustrate these ideas with the down-and-out call option in the following example.

Example 9.6   *Suppose that $S(0) = 100, r = 0.05, \sigma = 0.4, and \ T = 1$ (1 year). Estimate the delta of down-and-out call option with a strike price of 95 and provision on a downside barrier of 80.*

We base our simulation on 100,000 sample paths, each of which is divided into 100 equally spaced intervals. Therefore, this simulation requires 10 million independent normal random variables, namely ϵ_{ij} with $i = 1, 2, \dots, 100$ and $j = 1, 2, \dots, 100,000$. Using the set of $\{\epsilon_{ij}\}$, we produce the sample paths as $\{S_j(t_i), \dots, S_j(t_{100})\}$ using the Black–Scholes dynamics of asset price with $S_j(0) = 100$ for all j. Therefore, we get the C_{do} price as in Section 9.2. To obtain delta, we repeat the aforementioned procedure by assuming $S_j(0) = 100 + h$, where $h = 0.01$, to estimate the option price again. It is important to recall that we must use the same set of ϵ_{ij}. After that, the delta is approximated by the FDM. Our simulation estimates the delta of the down-and-out call option to be 0.863. Figure 9.3 shows the distribution of the delta estimates over 100 simulations. Please see the online material for the VBA codes.

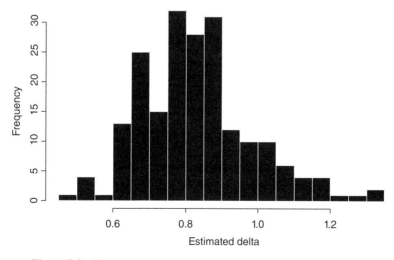

Figure 9.3   The strike against the delta of a down-and-out call option.

Other Greek letters can be obtained in a similar manner. For instance, the Gamma is the second order partial differentiation of the option pricing formula with respect to the underlying asset price. To estimate its value, we can approximate the second order differentiation by central finite differencing such that

$$\text{Gamma} \simeq \frac{V(S+h) - 2V(S) + V(S-h)}{h^2}.$$

Therefore, we are required to compute $V(S-h)$ on top of $V(S)$ and $V(S+h)$.

Example 9.7   *Using the input parameters in Example 9.6, plot the gamma of down-and-out call option against strike price, where the strike price varies from 93 to 110. Please see the online material for the VBA codes.*

9.7   EXERCISES

1. Verify Equations 9.1 and 9.3.

2. By modifying Example 9.1, simulate the price of down-and-out call, which will be knocked out if the underlying asset price goes below $8.

3. By modifying Example 9.1, simulate prices of a fixed lookback put option and a floating lookback call if the fixed strike price and the realized minimum asset prices are both $8. Verify the lookback put-call parities of these options.

4. Show that American call option price equals to that of its European counterpart if the underlying asset pays no dividends. In other words, American call option is never optimal to exercise before maturity if the underlying asset pays no dividends.

5. By modifying Example 9.4, simulate the price of an American call option with strike of $12 and a dividend yield δ of 4%. Hint: The risk-neutral dynamics of an asset paying continuous dividend yield is given by

$$\frac{dS}{S} = (r - \delta)\, dt + \sigma\, dW.$$

What is the optimal exercising policy from your simulation? Plot the critical asset prices against time.

6. Forward start option is a path dependent option that the strike price will be set as the underlying asset price in the future. For instance, the forward start call option payoff is

$$\max(S_T - S_{t_1}, 0),$$

where $0 < t_1 < T$.
   (a) Suppose $S_0 = \$10$, $\sigma = 0.4$, $r = 0.1$, $\delta = 0.5$, $T = 0.5$ and $t_1 = 0.3$. Construct and implement an algorithm for the forward start call option with 1,000 sample paths.

(b) Denote $C_{BS}(S, t; K, T)$ by the Black–Scholes formula for the standard call option. On the basis of financial insights, a risk analyst speculates that the forward start call option is the discounted standard call price. That is

$$\text{Current forward start call price} = e^{-rt_1} C_{BS}(S_0, t_1; S_0, T).$$

Verify this conjecture by your simulation.

(c) Suppose that the option has a continuous early exercise right after $t = t_1$. Determine the option price by the least squares simulation with 10,000 sample paths.

The solutions and/or additional exercises are available online at http://www.sta. cuhk.edu.hk/Book/SRMS/.

10

MULTIASSET OPTIONS

10.1 INTRODUCTION

Multiasset options are exotic options whose payoffs depend on values of multiple assets. Multiasset options abound in the financial market. An obvious example is index options, where the underlying variable, the financial index, can be thought of as a portfolio of multiple assets. Challenges of valuing multiasset options are the curse of dimensionality and the lack of analytical tractability. These problems can be circumvented by simulations.

Some examples of multiasset options traded in the financial market are first introduced. Let $S_1, S_2, \ldots, S_n$ denote the prices of n different assets.

1. *Exchange Options*. The right to exchange an asset for another. Thus, the option payoff is $\max(S_1 - cS_2, 0)$, where c is a constant multiplicative factor. This option is useful, for example, when a U.S. investor wants to buy Japanese yen with eurodollars.

2. *Quanto Options*. Options on stocks in a foreign country, that is, involving the exchange rate. If we treat S_1 as the exchange rate and S_2 as the underlying asset in the foreign country, then there are a number of possible quanto option payoffs, such as $S_1 \max(S_2 - K, 0), \max(S_1 S_2 - K, 0), \max(S_1, C), \max(S_2 - K, 0)$, and $C \max(S_2 - K, 0)$, where C is a fixed constant. The last payoff function appears to be of a single asset option. However, the volatility of the exchange rate, S_1, does contribute to the option price if S_1 and S_2 are correlated.

Simulation Techniques in Financial Risk Management, Second Edition. Ngai Hang Chan and Hoi Ying Wong.
© 2015 John Wiley & Sons, Inc. Published 2015 by John Wiley & Sons, Inc.

3. *Basket Options.* Options S on a portfolio. The payoff of a call on a portfolio is $\max(\Pi - K, 0)$, where $\Pi = \sum_{i=1}^{n} a_i S_i$.

4. *Extreme Options.* Options on the extrema of different assets. The maximum call option has the payoff: $\max\left[\max(S_1, S_2, \ldots, S_n) - K, 0\right]$.

All multiasset options can be traded with European or American style. Complex multiasset options, or structured products, may even involve path-dependent features. In such cases, simulations are indispensable.

10.2 SIMULATING EUROPEAN MULTIASSET OPTIONS

Consider an option on two assets with payoff $F(S_1(T), S_2(T))$. In the risk-neutral world, assets are assumed to follow the dynamics of

$$\frac{dS_i}{S_i} = r\, dt + \sigma_i\, dW_i, \quad i = 1, 2, \tag{10.1}$$

where

$$\widehat{E}(dW_1 dW_2) = \rho\, dt, \tag{10.2}$$

and $\widehat{E}$ denotes the risk-neutral expectation. Then, the option can be simulated via the Cholesky decomposition (Theorem 6.4).

Example 10.1 *Suppose that $S_1(0) = S_2(0) = 10$, $\sigma_1 = 0.3$, $\sigma_2 = 0.4$, $\rho = 0.2$, and $r = 0.05$. Simulate the price of an exchange option with maturity of 6 months.*

By Itô's lemma, we derive the terminal asset prices as

$$S_1(T) = S_1(0)e^{(r-\sigma_1^2/2)T + \sigma_1 X_1 \sqrt{T}} \quad \text{and} \quad S_2(T) = S_2(0)e^{(r-\sigma_2^2/2)T + \sigma_2 X_2 \sqrt{T}}, \tag{10.3}$$

where

$$\begin{bmatrix} X_1 \\ X_2 \end{bmatrix} \sim N\left(\begin{bmatrix} 0 \\ 0 \end{bmatrix}, \begin{bmatrix} 1 & \rho \\ \rho & 1 \end{bmatrix} \right).$$

The option price, C_X, can be determined by evaluating the expectation:

$$C_X = e^{-rT}\widehat{E}\left[\max(S_1(T) - S_2(T), 0)\right].$$

We estimate the option price by the following simulation algorithm.

Step 1: For $i = 1$ to N, perform Steps 2–4 as follows:

Step 2: Generate $Z_1, Z_2 \sim N(0,1)$ i.i.d. (identical and independent distributed)

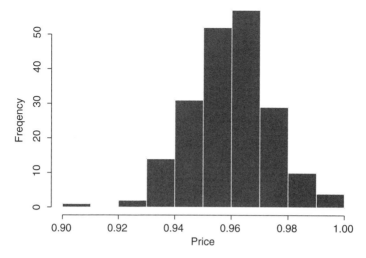

Figure 10.1 The distribution of simulated price.

Step 3: Set $X_1 = Z_1$ and $X_2 = \rho Z_1 + \sqrt{1 - \rho^2} Z_2$.

Step 4: Compute $S_1^{(i)}(T)$ and $S_2^{(i)}(T)$ by Equation 10.3.

Step 5: Set $C_X = \frac{e^{-rT}}{N} \sum_{i=1}^{N} \max(S_1^{(i)}(T) - S_2^{(i)}(T), 0)$.

Figure 10.1 plots the distribution of the estimated price over 100 simulations. We obtain the estimated option price to be 0.962.

10.3 CASE STUDY: ON ESTIMATING BASKET OPTIONS

In practice, basket options are often valued by assuming that the value of the portfolio of assets comprising the basket follows the Black–Scholes dynamics jointly rather than that each asset follows the Black–Scholes dynamics individually. After estimating the portfolio volatility from the portfolio return, the basket call option is valued by substituting the portfolio volatility into the Black–Scholes formula. This approach offers a quick solution to traders. However, the risk manager needs to understand the risk of this simplifying assumption. We examine this approach by means of simulation.

Consider a basket call option with three underlying assets, S_1, S_2, and S_3. The payoff of this option is $\max(S_1 + S_2 + S_3 - K, 0)$. In other words, the holder of the option has the right to purchase the portfolio as a sum of the three assets for a fixed value of K. Suppose that the current time is $t = 1$, and we observe the prices of three assets since $t = 0$. Figure 10.2 depicts the paths of the three simulated asset prices.

At $t = 1$, the asset prices are $S_1 = 142.69$, $S_2 = 89.23$, and $S_3 = 49.73$. The current portfolio value is the sum of three assets and equals 281.65. On the basis of the three

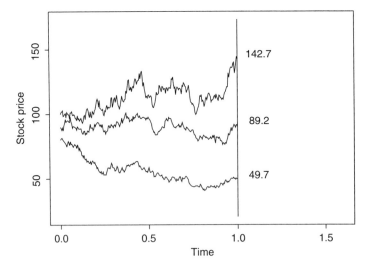

Figure 10.2 The historical price of shocks.

asset price paths, the portfolio volatility is estimated to be 0.280. Consider the basket option with a strike price of 250, maturity of half a year, and interest rate of 5%. The naive application of the Black–Scholes formula produces a value for the option as 44.81.

On the other hand, we can use MC (Monte Carlo) simulation to estimate the option price by assuming individual asset follows the Black–Scholes dynamics. By examining the asset price paths, we estimate the variance–covariance matrix for assets returns as

$$\begin{pmatrix} 0.172 & 0.050 & 0.043 \\ 0.050 & 0.088 & 0.038 \\ 0.043 & 0.038 & 0.123 \end{pmatrix}.$$

Then, we simulate asset prices at $t = 1.5$ using the Cholesky decomposition for 10,000 times. Figure 10.3 illustrates the idea of generating asset values at $t = 1.5$, the maturity of the option. Terminal values of individual assets are simulated on the basis of an approach similar to Equation 10.3 with three assets. The option price is then evaluated by discounting the sample mean of the option payoff using the interest rate of 5%. The simulated option price is 51.35, which is larger than the naive approach of 44.81.

We are also interested in the contribution of the error in estimating parameters of the option. We perform a control experiment assuming that the variance–covariance matrix can be estimated without error. Input the variance–covariance matrix as

$$\begin{pmatrix} 0.1600 & 0.0360 & 0.0420 \\ 0.0360 & 0.0900 & 0.0315 \\ 0.0420 & 0.0315 & 0.1225 \end{pmatrix}.$$

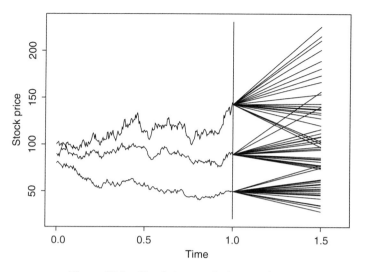

Figure 10.3   Simulating terminal asset prices.

Using the same set of independent normal random numbers, we obtain the option price as 50.71. It appears that the error of estimating the variance–covariance matrix does not contribute too much to basket option values. Therefore, the MC and the naive approach to valuing basket options can produce significantly different results, irrespective of the estimation error.

In practice, banks and financial institutions usually have a lot of derivatives positions in their portfolio. Risk managers are responsible to check for the consistency of models that are used to value individual derivatives in the portfolio. Imagine a situation that a bank buys and sells options on individual assets and a basket of assets everyday. When individual assets are assumed to follow the Black–Scholes dynamics, it is crucial for the risk manager to realize what kind of assumptions have been imposed. The simulation shows that it is not appropriate to assume the portfolio constituting the basket to follow Black–Scholes dynamics jointly because this assumption is not consistent with the assumption on individual assets. In such a case, the value of basket options can be significantly underestimated.

10.4 DIMENSION REDUCTION

For an n-asset option, simulation can be constructed by using the Cholesky decomposition (Eq. 6.6). However, this requires generating n independent normal random variables for each scenario. To reduce the computational burden, we can use the principle component analysis (PCA) to approximate the n factors by a smaller number of factors, usually less than 10 in practice.

Suppose that we have an n dimensional random vector $X \sim N(0, \Sigma)$ where Σ is an $n \times n$ variance–covariance matrix. PCA for normal random variables is to approximate X by Y, which follows a distribution similar to that of X but is easier to simulate.

PCA uses the eigenvalue decomposition in Chapter 6 to approximate the random vector X. Let $\lambda_1, \lambda_2, \ldots, \lambda_n$ be eigenvalues of Σ and $v_1, v_2, \ldots, v_n$ be the corresponding eigenvectors. As variance–covariance matrices are positive definite, their eigenvalues are all positive real numbers so that the corresponding squared roots are positive real numbers. Theorem 6.6 asserts that the random vector

$$X = \sqrt{\lambda_1} v_1 Z_1 + \sqrt{\lambda_2} v_2 Z_2 + \cdots + \sqrt{\lambda_n} v_n Z_n, \qquad (10.4)$$

where $Z_1, Z_2, \ldots, Z_n$ are i.i.d. standard normal random variables. The equality (Eq. 10.4) is defined in the sense of distribution. In PCA, we arrange eigenvalues in descending order such that $\lambda_1 > \lambda_2 > \cdots > \lambda_n$. From Equation 10.4, we see that the contribution of the term $\sqrt{\lambda_i} v_i Z_i$ to the value of X decreases with the index i. The eigenvector v_i is called the ith principle component (PC). To approximate X, we truncate the sum in Equation 10.4 such that

$$X \simeq \sqrt{\lambda_1} v_1 Z_1 + \sqrt{\lambda_2} v_2 Z_2 + \cdots + \sqrt{\lambda_m} v_m Z_m,$$

where $m < n$. If we are comfortable with this approximation, we then simulate m independent standard normal random variables Z_i and calculate everything on the basis of this approximation.

An important topic in PCA is to determine the value m. The number of terms used in the approximation depends on the accuracy of the outcome required by the modeler. If the user requires 100% accuracy besides simulation error, then he should use formula of Equation 10.4. PCA is useful when the user requires an accuracy that is less than 100%. Suppose he requires an accuracy of at least 99%. Then, m is the minimum integer such that

$$\frac{\sum_{i=1}^{m} \lambda_i}{\sum_{i=1}^{n} \lambda_i} \geq 99\%.$$

A proof of this result can be found in standard texts in multivariate analysis, for example Anderson (2003).

Let us apply PCA in multiasset option pricing. Consider an option with 10 underlying assets. Each asset follows the Black–Scholes dynamics such that

$$dS_i = \mu_i S_i \, dt + \sigma_i S_i \, dW_i, \quad i = 1, 2, \ldots, 10,$$

where S_i is the value of the ith asset. The $W_1, W_2, \ldots, W_{10}$ are correlated Brownian motions with correlation matrix:

$$\begin{pmatrix}
1.00 & 0.74 & 0.34 & -0.08 & 0.05 & -0.74 & 0.04 & -0.12 & 0.81 & 0.82 \\
0.74 & 1.00 & 0.81 & -0.04 & -0.57 & -0.25 & 0.06 & 0.47 & 0.89 & 0.92 \\
0.34 & 0.81 & 1.00 & -0.17 & -0.83 & 0.20 & -0.09 & 0.78 & 0.65 & 0.72 \\
-0.08 & -0.04 & -0.17 & 1.00 & 0.01 & -0.05 & 0.94 & -0.04 & -0.09 & -0.05 \\
0.05 & -0.57 & -0.83 & 0.01 & 1.00 & -0.55 & 0.00 & -0.94 & -0.41 & -0.45 \\
-0.74 & -0.25 & 0.20 & -0.05 & -0.55 & 1.00 & -0.16 & 0.65 & -0.40 & -0.40 \\
0.04 & 0.06 & -0.09 & 0.94 & 0.00 & -0.16 & 1.00 & -0.06 & 0.04 & 0.06 \\
-0.12 & 0.47 & 0.78 & -0.04 & -0.94 & 0.65 & -0.06 & 1.00 & 0.31 & 0.34 \\
0.81 & 0.89 & 0.65 & -0.09 & -0.41 & -0.40 & 0.04 & 0.31 & 1.00 & 0.91 \\
0.82 & 0.92 & 0.72 & -0.05 & -0.45 & -0.40 & 0.06 & 0.34 & 0.91 & 1.00
\end{pmatrix}.$$

A discrete approximation to the asset price dynamics is

$$\Delta S_i = r S_i \,\Delta t + S_i \sqrt{\Delta t}\, \epsilon_i,$$

where ϵ_i are risk factors such that $[\epsilon_i] = X \sim N(0, \Sigma)$, and Σ is the correlation matrix given previously.

For the given correlation matrix, eigenvalues are obtained as 4.719, 2.843, 1.931, 0.147, 0.104, 0.079, 0.062, 0.056, 0.038, 0.022. Summing up all the eigenvalues gives a value of 10. When we divide the sum of the first three eigenvalues by the total sum, the ratio is close to 95%. Therefore, if we accept an error of 5%, the first three PCs provide sufficient accuracy. Eigenvectors corresponding to the first three PCs are found to be:

$$
\begin{array}{lrrrrrrrrrr}
v_1: & 0.31 & 0.44 & 0.41 & -0.05 & -0.31 & -0.07 & -0.00 & 0.27 & 0.42 & 0.43 \\
v_2: & 0.41 & 0.08 & -0.22 & 0.09 & 0.41 & -0.57 & 0.14 & -0.45 & 0.18 & 0.16 \\
v_3: & 0.09 & -0.03 & 0.01 & -0.70 & 0.12 & -0.05 & -0.69 & -0.10 & 0.02 & -0.00
\end{array}
$$

On the basis of the first three PCs, we generate three independent standard normal random variables, namely Z_1, Z_2, and Z_3, and approximate the n risk factors by

$$[\epsilon_i] \simeq Z_1 \sqrt{\lambda_1}\, v_1 + Z_2 \sqrt{\lambda_2}\, v_2 + Z_3 \sqrt{\lambda_3}\, v_3.$$

The 10 risk factors $\epsilon_1, \epsilon_2, \dots, \epsilon_{10}$ are reduced to only three independent factors. Hence, we reduce a 10-dimensional problem to a three-dimensional problem.

Example 10.2 *Value a maximum option on 10 assets with a strike price of $95 and a maturity of half a year. All asset values are currently $100 with volatilities of 30% for all assets. The correlation matrix of risk factors is given previously. The interest rate is 4%. We accept a maximum error of 5%.*

The option payoff is $\max[\max(S_1, S_2, \dots, S_{10}) - 95, 0]$. As the option is traded in European style, it is efficient to simulate terminal asset values directly. By Itô's lemma, we know that the terminal value of the ith asset is given by

$$
\begin{aligned}
S_i(T) &= S_i(0) \exp\left[(r - \sigma_i^2/2)T + \sigma_i W_i(T)\right] \\
&= S_i(0) \exp\left[(r - \sigma_i^2/2)T + \sigma_i \sqrt{T}\, \epsilon_i\right],
\end{aligned}
\tag{10.5}
$$

where the vector $[\epsilon_i] \sim N(0, \Sigma)$. Our simulation obtains the option price to be 49.12. See the online material for the VBA codes.

10.5 EXERCISES

1. Suppose that $x(t)$ and $y(t)$ are two correlated Itô's processes such that

$$
\begin{aligned}
dx &= a(t, x)\, dt + b(t, x)\, dW_1, \\
dy &= \alpha(t, y)\, dt + \beta(t, y)\, dW_2, \\
E(dW_1 dW_2) &= \rho\, dt.
\end{aligned}
$$

Consider a function, $f(t, x, y)$, which depends on both stochastic variables of $x(t)$ and $y(t)$. By modifying the proof of Theorem 4.1, show that the dynamic of $f(t, x, y)$ is

$$
\begin{aligned}
df &= \left(\frac{\partial f}{\partial t} + a\frac{\partial f}{\partial x} + \alpha\frac{\partial f}{\partial y} + \frac{b^2}{2}\frac{\partial^2 f}{\partial x^2} + \frac{\beta^2}{2}\frac{\partial^2 f}{\partial y^2} + \rho b\beta\frac{\partial^2 f}{\partial x\partial y} \right) dt \\
&\quad + b\frac{\partial f}{\partial x}\, dW_1 + \beta\frac{\partial f}{\partial y}\, dW_2.
\end{aligned}
\tag{10.6}
$$

This formula is known as the Itô's lemma for two variables.

2. Answer the following questions by considering the property of martingales defined in Question 7 of Chapter 5.

   (a) Consider a pair of asset price dynamics under the risk-neutral measure:

   $$
   \begin{aligned}
   dS_1 &= rS_1\, dt + \sigma_1 S_1\, dW_1, \\
   dS_2 &= rS_2\, dt + \sigma_2 S_2\, dW_2, \\
   E(dW_1 dW_2) &= \rho\, dt.
   \end{aligned}
   $$

   Show that the stochastic process $X(t) = S_1(t)/S_2(t)$ is a martingale under the Brownian motions $W_1^*(t)$ and $W_2^*(t)$ where

   $$
   W_1^*(t) = W_1(t) - \rho\sigma_2 t \quad \text{and} \quad W_2^*(t) = W_2(t) - \sigma_2 t.
   $$

   (b) Under (a), show that $X(t)$ has the dynamics:

   $$
   \frac{dX}{X} = \sigma\, dW^*,
   $$

   where W^* is a Brownian motion and

   $$
   \sigma^2 = \sigma_1^2 - 2\rho\sigma_1\sigma_2 + \sigma_2^2.
   $$

(c) Consider a function of S_1 and S_2, $V(t, S_1, S_2)$, which has the property that

$$V(t, S_1, S_2) = S_2\, U(t, S_1/S_2) = S_2\, U(t, X).$$

Show that $U(t, X)$ is a martingale under Brownian motions $W_1^*(t)$ and $W_2^*(t)$.

3. Consider the exchange option with payoff $\max(S_1(T) - S_2(T), 0)$. Denote the option pricing formula for this option as $V_{ex}(t, S_1, S_2)$. By using the no-arbitrage argument, one derives that the exchange option has the properties:

- There exists a function U such that $V_{ex}(t, S_1, S_2) = S_2\, U(t, S_1/S_2)$.
- There exists a probability measure $\mathcal{Q}$ such that $X(t) = S_1(t)/S_2(t)$ is a martingale.

On the basis of these properties and the results obtained in Question 2, show that

$$V_{ex} = S_1 \Phi(d_1^*) - S_2 \Phi(d_2^*),$$

where

$$d_1^* = \frac{\log(S_1/S_2) + \sigma^2(T - t)/2}{\sigma\sqrt{T - t}},$$
$$d_2^* = d_1^* - \sigma\sqrt{T - t},$$
$$\sigma^2 = \sigma_1^2 - \rho\sigma_1\sigma_2 + \sigma_2^2.$$

Herein, σ_1 and σ_2 are volatilities of S_1 and S_2, respectively, and ρ is the correlation coefficient between the returns of two assets. This formula was first derived in Margrabe (1978).

4. Run the simulation program for pricing exchange option and compare the numerical result with the analytical one.

5. The so-called geometric basket option has the payoff function

$$\max\left\{\left(\prod_{i=1}^{n} S_i(T)\right)^{1/n} - K, 0\right\}.$$

(a) Show that this option has a value less than the usual basket option with payoff

$$\max\left\{\frac{1}{n}\sum_{i=1}^{n} S_i(T) - K, 0\right\}.$$

(b) Suppose that individual assets follow the Black–Scholes dynamics. Derive the analytical pricing formula for the geometric basket option.

(c) By regarding the price of the geometric basket option as a control variate, simulate the price of the usual basket option that depends on four assets with the following correlation matrix:

$$\begin{pmatrix} 1.0000 & 0 & 0.3000 & 0.3000 \\ 0 & 1.0000 & 0.4000 & 0.2000 \\ 0.3000 & 0.4000 & 1.0000 & 0.3000 \\ 0.3000 & 0.2000 & 0.3000 & 1.0000 \end{pmatrix}.$$

We assume all assets sharing the same volatility of 30%, and each asset individually follows the Black–Scholes dynamics.

6. Use simulation to determine the value and early exercise policy of American style exchange options. We assume the interest rate of 5%, $S_1 = 100$, $S_2 = 95$, $T = 1$ year, and the variance–covariance matrix of asset returns:

$$\begin{pmatrix} 0.016 & 0.006 \\ 0.006 & 0.09 \end{pmatrix}.$$

Hint: you may use the Least Squares model and a quadratic polynomial of S_1 and S_2 in your regression.

The solutions and/or additional exercises are available online at http://www.sta.cuhk.edu.hk/Book/SRMS/.

11

INTEREST RATE MODELS

11.1 INTRODUCTION

Fixed income securities are concerned with the valuation of promised payments at a future date. For example, a zero coupon bond promises to pay a single payment on the maturity day. A straight U.S. Treasury bond promises to make payments, the amount and date of which are determined by the face value, maturity date, and coupon rate of the bond. Because cash flows are certain, we are not concerned with the risk of the volatility of the amount of cash. Instead, we are interested in the following question: *How much would a rational individual be willing to pay today for a promised payment in the future?* The answer to this question is related to the movement of the interest rate, which leads to the next question: *What is the best way to manage the interest rate risk?* Simulation can serve as a useful tool in answering these questions.

11.2 DISCOUNT FACTOR AND BOND PRICES

Consider the simplest case in which a zero coupon bond (zero) will pay $1 a year from now. What is the maximum that one should be willing to pay for this contract today? Purchasing this bond should be worth at least as much as putting the money in the bank. Let P be the payment at the current moment. Then,

$$P(1 + R) = 1,$$

Simulation Techniques in Financial Risk Management, Second Edition. Ngai Hang Chan and Hoi Ying Wong.
© 2015 John Wiley & Sons, Inc. Published 2015 by John Wiley & Sons, Inc.

where R is the current annual interest paid by a bank (R is supposed to be a constant). That is,

$$P = \frac{1}{1 + R}.$$

P and $\frac{1}{1+R}$ are known as the zero price and the discount factor, respectively.

Now we define $P(t, T)$ as the zero coupon bond price at time t with maturity at time T. A typical bond will pay coupons at semiannual intervals and a principle payment at maturity. Figure 11.1 illustrates the cash flow of a 3-year coupon bond. The key to evaluating such bond is to view the amounts promised at different future dates as separate zero coupon bonds. We then value each payment at each date using the discount factor for that date and sum up the values. Let $\widetilde{P}(t, T)$ be the corresponding coupon-bearing bond price with a coupon rate $C(t_i)$ paid at each coupon payment date t_i, $i = 1, \ldots, N$. Then, $\widetilde{P}(t, T)$ can be valued by the formula:

$$\widetilde{P}(0, T) = \sum_{i=1}^{N} C(t_i) P(0, t_i). \tag{11.1}$$

For example, the value of a bond paying a semiannual coupon is given by

$$\widetilde{P}(0, T) = \frac{C(1/2)}{1 + R/2} + \frac{C(1)}{(1 + R/2)^2} + \cdots + \frac{C(N) + 1}{(1 + R/2)^{2N}}.$$

To simplify the mathematics, we define r as the continuously compounded interest rate. Its relationship with the annual interest rate R is given by the formula

$$\frac{1}{1 + R} = e^{-r}.$$

In reality, interest rates are not constants but change over time. From now on, we assume that the continuously compounding interest rate r is a function of time t, that is, $r = r_t$, and we call it the instantaneous interest rate. Suppose that we invest \$1 in the money market account $B(0)$ today with the interest rate r_t, then the interest will

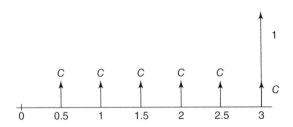

Figure 11.1   Cash flow of a 3-year coupon bond.

rollover continuously at every instant by $B(t)r_t\,dt$. At any time t, the money market account $B(t)$ satisfies

$$dB(t) = B(t)r_t\,dt.$$

Solving the aforementioned differential equation, with the initial condition $B(0) = 1$, we obtain

$$B(t) = e^{\int_0^t r_s\,ds}. \tag{11.2}$$

Conversely, if a bond pays the holder \$1 at future time t, the bond is worth $1/B(t) = e^{-\int_0^t r_s ds}$ dollars at time 0, and we will use it as the discount factor for future cash flow. So, for a deterministic interest rate r_t, the time t zero coupon bond price $P(t,T)$ with maturity at T is given by

$$P(t,T) = e^{-\int_t^T r_s\,ds}. \tag{11.3}$$

Clearly, $P(T,T) = 1$. In the following section, we extend the interest rates to be stochastic. Practically, the zero coupon bond price is expressed in terms of the continuous yield to maturity $R(t,T)$ by

$$P(t,T) = e^{-R(t,T)(T-t)}. \tag{11.4}$$

This yield to maturity corresponds to the constant interest rate of the continuously compounded interest rate from time t to T and can serve as an indicator of the price of the bond. If we are given the zero coupon bond prices from the market, the yield can be recovered by

$$R(t,T) = -\frac{\log P(t,T)}{T-t}. \tag{11.5}$$

Similarly, for a coupon-bearing bond, $\widetilde{P}(0,T)$ and $R(0,T)$ are related by

$$\widetilde{P}(0,T) = C(t_1)e^{-R(0,t_1)t_1} + C(t_2)e^{-R(0,t_2)t_2} + \cdots + (C(t_N) + 1)e^{-R(0,t_N)t_N},$$

where $C(t_i)$ is the coupon paid at time t_i, $i = 1, \ldots, N$.

Bond markets usually quote the yield in place of the interest rate r_t. Bond prices are available only for some discrete times to maturity T_i, $i = 1, \ldots, N$, such as 1-, 3-, and 5-year, so it is more convenient if we parametrize $R(t,T)$ as a piecewise continuous function and interpolate all of the discrete points to obtain a continuum of $R(t,T)$ for all $T \geq 0$.

For example, $R(0,T)$ can be parametrized by a piecewise smooth cubic function as follows:

$$R(0,T) = \begin{cases} a_0 + b_0 T + c_0 T^2 + d_0 T^3, & \text{for } T \in [0, T_0], \\ a_1 + b_1 T + c_1 T^2 + d_1 T^3, & \text{for } T \in [T_0, T_1]. \end{cases} \tag{11.6}$$

We can then use interpolation methods, such as cubic spline, to find the coefficients by putting the market bond data into formula 11.5. Further discussions about yields and interest rate models can be found in Jarrow (2002).

Example 11.1 *Suppose $R(0,T)$ is parametrized as*

$$R(0,T) = \begin{cases} 0.005 - 0.001T - 0.0001T^2 + 0.0005T^3, & \text{for } T \in [0,2], \\ 0.0078 - 0.0052T + 0.002T^2 + 0.00015T^3, & \text{for } T \in [2,3], \end{cases}$$

then it is first-order continuous. Figure 11.2 shows the graph of $R(0,T)$ in Excel for $0 \le T \le 3$. The corresponding discount curve $P(0,T)$ can also be obtained easily from Figure 11.3.

For a 3-year coupon bond with a notional value of $100 and with a coupon rate of 6% paid semiannually, the price is given by

$$\widetilde{P}(0,3) = 100\left[0.03e^{-R(0,0.5)0.5} + 0.03e^{-R(0,1)1} + \cdots + (0.03+1)e^{-R(0,3)3}\right]$$

$$= 113.537.$$

This Excel file can be downloaded online.

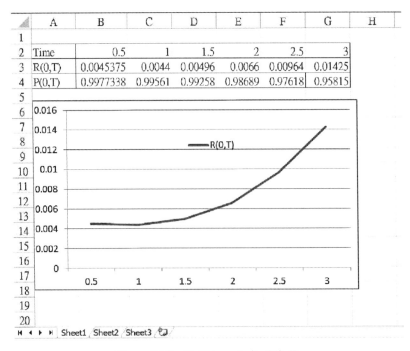

Figure 11.2 Yield to maturity $R(0,T)$.

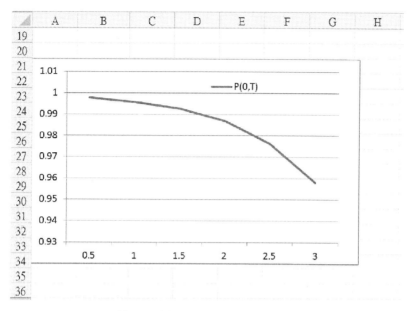

Figure 11.3   Discount curve $P(0, T)$.

11.3   STOCHASTIC INTEREST RATE MODELS AND THEIR SIMULATIONS

Deterministic interest rate models are inadequate for capturing interest rate movements, as the future interest rates cannot be known for certain. A better approach is to incorporate the stochastic feature of the interest rates. A stochastic interest rate model should match Equation 11.3 when the stochastic component is absent. A natural way is to consider

$$P(t, T) = \mathrm{E}\left[e^{-\int_t^T r(s, \mathbf{W}_s)\,ds} \middle| \mathcal{F}_t\right],$$ (11.7)

where $\mathbf{W}_s$ is a vector of stochastic factors, and $\mathcal{F}_t$ is the filtration generated by $\{\mathbf{W}_s, t \geq s \geq 0\}$. Intuitively, $\mathcal{F}_t$ consists of all of the information available up to time t. Given that at current time t, the future interest rates from time t to T are all random, we need to take the expectation conditional on the current information. If stochastic factors are absent, the function inside the expectation becomes deterministic and the expectation is equal to the function itself.

For pricing derivatives with stochastic interest rates, the future cash flows are discounted using the zero coupon bond price from Equation 11.7 under the risk neutral expectation. There are many ways to model the interest rate movement. In this case, we consider the short rate models, in which the instantaneous interest rate r_t is specified by a stochastic differential equation (SDE).

From a simulation perspective, expression 11.7 offers a means to conduct Monte Carlo simulations. Once an appropriate stochastic interest rate model, such as the Vasicek model of Vasicek (1977), the CIR model of Cox, Ingersoll and Ross (1985), the Ho–Lee model of Ho and Lee (1986), and the Hull–White model of Hull and White (1988), is formulated, simulations can be conducted.

To illustrate this idea, consider a short rate model that follows

$$dr_t = \mu(t, r_t)\, dt + \beta(t, r_t)\, dW_t, \tag{11.8}$$

where r_t is the current continuously compounded interest rate and W_t is a Wiener process. For example, the Vasicek model assumes that $\mu(t, r) = a(b - r)$ and $\beta(t, r) = \sigma$, whereas the CIR model uses the same $\mu(t, r)$ with $\beta(t, r) = \sigma\sqrt{r}$. Sample paths for short rate models in the form of Equation 11.8 can be generated by the following steps:

Step 1: Set $r_i = r_0$ to be the current market rate.
Step 2: Generate $\epsilon \sim N(0, 1)$.
Step 3: Set $r_{i+1} = r_i + \mu(t_i, r_i)\,\Delta t + \beta(t_i, r_i)\,\epsilon\sqrt{\Delta t}$.
Step 4: Go to Step 2.

Let $r^{(j)} = \{r^{(j)}(t) : t = 0, \frac{1}{n}, \frac{2}{n}, \ldots, T\}$ be the j-th interest rate path out of M sample paths generated by the preceding algorithm with $\Delta t = \frac{1}{n}$. By means of quadrature, we can make the following approximation:

$$\int_{t_1}^{t_2} r^{(j)}(t)\, dt \simeq \frac{1}{n} \sum_{t \in [t_1, t_2]} r^{(j)}(t). \tag{11.9}$$

If we take $\Delta t = \frac{1}{280}$, the zero coupon bond price $P(0, 1)$ can approximated by

$$
\begin{aligned}
P(0, 1) &\simeq E\left\{ \exp\left(-\frac{1}{280} \sum_{i=0}^{280} r^{(j)}\left(\frac{i}{280}\right) \right) \right\} \\
&\simeq \frac{1}{M} \sum_{j=1}^{M} \exp\left(-\frac{1}{280} \sum_{i=0}^{280} r^{(j)}\left(\frac{i}{280}\right) \right).
\end{aligned}
$$

In general, we write

$$
\begin{aligned}
P(t, T) &\simeq \frac{1}{M} \sum_{j=1}^{M} \exp\left(-\frac{1}{n} \sum_{t_i \in [t, T]} r^{(j)}(t_i) \right) \\
&= \frac{1}{M} \sum_{j=1}^{M} \exp\left(-\mathrm{Avg}_{t_i \in [t, T]} r^{(j)}(t_i) \times (T - t) \right). \tag{11.10}
\end{aligned}
$$

11.4   HULL–WHITE MODEL

Although many short rate models have been proposed to model the dynamics of interest rates, we illustrate the pricing of zero coupon bonds and calibration of the model parameters under the Hull and White (1994) model. This model admits the analytical bond price formula and can therefore simplify the pricing of other exotic fixed income derivatives. The instantaneous interest rate r_t is assumed to follow the SDE, as follows:

$$dr_t = [\theta(t) - ar_t]\,dt + \sigma\,dW_t, \tag{11.11}$$

where $\theta(t)$ is a deterministic function of time, a and σ are constants, and W_t is a Wiener process. Applying Itô's lemma to $e^{at}r_t$, we have

$$d(e^{at}r_t) = \theta(t)e^{at}\,dt + e^{at}\sigma\,dW_t.$$

Rewriting the aforementioned equation into the integral form with some simplifications yields the following representation of r:

$$r_T = r_t e^{-a(T-t)} + \int_t^T \theta(\tau)e^{-a(T-\tau)}\,d\tau + \sigma\int_t^T e^{-a(T-\tau)}\,dW_\tau. \tag{11.12}$$

The following fact is useful in interest rate modeling. For a deterministic function $y(t)$, let

$$I(t) = \int_0^t y(s)\,dW_s,$$

then $I(t)$ is a normal random variate with a mean of 0 and variance $\int_0^t y^2(s)\,ds$. The proof is in Exercise 1. Therefore the interest rate r is normally distributed. By Itô's identities, see Exercise 1(d) in Chapter 4, the conditional expectation and variance of r_T given at time t are

$$E[r_T|r_t] = r_t e^{-a(T-t)} + \int_t^T \theta(\tau)e^{-a(T-\tau)}\,d\tau, \tag{11.13}$$

and

$$\mathrm{Var}(r_T|r_t) = E\left[\left(\sigma\int_t^T e^{-a(T-\tau)}\,dW_\tau\right)^2\right]$$

$$= \sigma^2\int_t^T e^{-2a(T-\tau)}\,d\tau$$

$$= \frac{\sigma^2}{2a}[1 - e^{-2a(T-t)}]. \tag{11.14}$$

To derive the analytical formula for a zero coupon bond, we need to evaluate $\int_t^T r_\tau d\tau$. By changing the order of integration,

$$\int_t^T r_\tau d\tau = \int_t^T r_t e^{-a(u-t)} du + \int_t^T \int_t^u \theta(\tau) e^{-a(u-\tau)} d\tau\, du$$

$$+ \int_t^T \int_t^u e^{-a(u-\tau)} dW_\tau\, du$$

$$= r_t \left(\frac{1 - e^{-a(T-t)}}{a} \right) + \int_t^T \int_\tau^T \theta(\tau) e^{-a(u-\tau)} du\, d\tau$$

$$+ \int_t^T \int_\tau^T e^{-a(u-\tau)} du\, dW_\tau.$$

$$= r_t \left(\frac{1 - e^{-a(T-t)}}{a} \right) + \int_t^T \theta(\tau) \left(\frac{1 - e^{-a(T-\tau)}}{a} \right) d\tau$$

$$+ \int_t^T \frac{\sigma}{a} [1 - e^{-a(T-\tau)}] dW_\tau. \qquad (11.15)$$

Therefore, $\int_t^T r_\tau d\tau$ is still normally distributed with the mean and variance given as

$$E\left[\int_t^T r_\tau d\tau \middle| r_t \right] = r_t \left(\frac{1 - e^{-a(T-t)}}{a} \right) + \int_t^T \theta(\tau) \left(\frac{1 - e^{-a(T-\tau)}}{a} \right) d\tau \quad (11.16)$$

and

$$\text{Var}\left(\int_t^T r_\tau d\tau \middle| r_t \right) = \int_t^T \frac{\sigma^2}{a^2} [1 - e^{-a(T-\tau)}]^2\, d\tau$$

$$= \frac{\sigma^2}{2a^3} \left[2a(T - t) - 3 + 4e^{-a(T-t)} - e^{-2a(T-t)} \right]. \qquad (11.17)$$

The moment-generating function of a normal random variable X is given by

$$E[e^{uX}] = \exp\left\{ uE[X] + \frac{u^2}{2} \text{Var}(X) \right\}.$$

The zero coupon price is

$$P(t, T) = E\left[e^{-\int_t^T r_\tau d\tau} \middle| \mathcal{F}_t \right]$$

$$= \exp\left\{ C(t, T) - D(t, T)r_t \right\}, \qquad (11.18)$$

where

$$D(t, T) = \left(\frac{1 - e^{-a(T-t)}}{a} \right),$$

$$C(t, T) = - \int_t^T \theta(\tau) \left(\frac{1 - e^{-a(T-\tau)}}{a} \right) d\tau$$

$$+ \frac{\sigma^2}{4a^3} \left[2a(T - t) - 3 + 4e^{-a(T-t)} - e^{-2a(T-t)} \right].$$

Another method of deriving the solution by the PDE approach is provided in the exercise. Now we assume that a and σ are known. If we want to use the model, then we have to calibrate $\theta(t)$ to the current market zero coupon bond prices. In other words, given $P(t, T)$ from the market data for different maturities T, we need to express $\theta(t)$ in terms of $P(t, T)$, which is more complicated. Therefore, it is more convenient to decompose r_t by

$$r_t = \alpha(t) + x_t, \tag{11.19}$$

and x_t follows

$$dx_t = -ax_t \, dt + \sigma \, dW_t.$$

$\alpha(t)$ is a deterministic function that incorporates the information in $\theta(t)$, and x_t corresponds to the random component driven by W_t. We take $x_0 = 0$ and $\alpha(0) = r_0$. To simulate r_t, we just need to perform the simulation on x_t and add the corresponding $\alpha(t)$ at each step. Figure 11.4 plots the graph of a simulated path of x_t and Figure 11.5 shows an example of $\alpha(t)$. Their sum leads to the sample path of r_t in Figure 11.6.

Note that x_t actually follows a degenerate Hull–White model; in fact, it is an Ornstein–Uhlenbeck process, where $\theta(t) \equiv 0$ for all t. This property will be useful

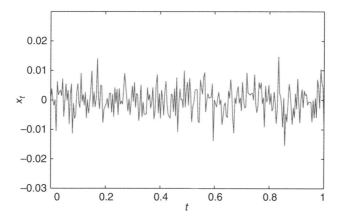

Figure 11.4 Simulated sample path of x_t.

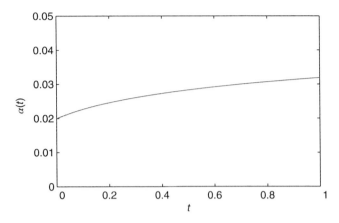

Figure 11.5 $\alpha(t)$.

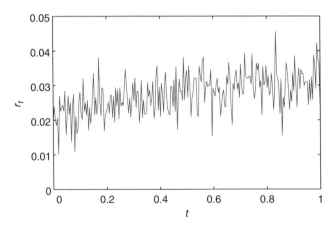

Figure 11.6 r_t.

for evaluating a similar expectation to that in Equation 11.18 in terms of x_t. Denote $P_0(t, T)$ and $C_0(t, T)$ as the corresponding $P(t, T)$ and $C(t, T)$ in Equation 11.18 when $\theta(t) \equiv 0$, respectively. To express $\alpha(t)$ in terms of $P(t, T)$, we evaluate

$$P(t, T) = \mathrm{E}\left[e^{-\int_t^T (\alpha(\tau) + x_\tau)\, d\tau}\middle| \mathcal{F}_t\right]$$

$$= e^{-\int_t^T \alpha(\tau) d\tau} \mathrm{E}\left[e^{-\int_t^T x_\tau\, d\tau}\middle| \mathcal{F}_t\right]$$

$$= e^{-\int_t^T \alpha(\tau)\, d\tau} P_0(t, T). \qquad (11.20)$$

Taking the natural logarithm on both side yields

$$\int_t^T \alpha(\tau)\, d\tau = -\log \frac{P(t,T)}{P_0(t,T)}.$$

Assume that $t = 0$, and differentiate both sides with respect to T, then we have

$$\alpha(T) = -\frac{\partial}{\partial T} \log \frac{P(0,T)}{P_0(0,T)}. \tag{11.21}$$

Example 11.2   *When $R(0,T)$ is estimated from the market data using the parametric form given in Equation 11.6, $\alpha(T)$ can be computed explicitly.*

Substitute $R(0,T)$ into formula 11.21 gives

$$\alpha(T)$$

$$= -\frac{\partial}{\partial T} \log \left[e^{-R(0,T)T - C_0(0,T) + D(0,T)x_0} \right]$$

$$= \begin{cases} a_0 + 2b_0 T + 3c_0 T^2 + 4d_0 T^3 + \dfrac{\sigma^2}{2a^2}(1 - 2e^{-aT} + e^{-2aT}), & \text{for } T \in [0, T_0], \\[2ex] a_1 + 2b_1 T + 3c_1 T^2 + 4d_1 T^3 + \dfrac{\sigma^2}{2a^2}(1 - 2e^{-aT} + e^{-2aT}), & \text{for } T \in [T_0, T_1]. \end{cases}$$

Knowing $\alpha(T)$ in closed form, the sample paths in the Hull–White model can be generated by the following steps:

Step 1: Set $r_i = r_0$ be the current market rate and $x_0 = 0$.
Step 2: Generate $\epsilon \sim N(0,1)$.
Step 3: Set $x_{i+1} = x_i - ax_i \Delta t + \sigma \epsilon \sqrt{\Delta t}$.
Step 4: Set $r_{i+1} = \alpha(t_{i+1}) + x_{i+1}$.
Step 5: Go to Step 2.

11.5   FIXED INCOME DERIVATIVES PRICING

For standard European options, introducing a stochastic interest rate into the Black-Scholes model has a minimal effect on prices, so interest rates are usually taken to be constant for vanilla stock options. However, for interest-rate-sensitive instruments, such as options on bond and range accrual notes, stochastic interest rates should be used in the model.

Consider a coupon bond selling at time T, assuming now is time 0, with a coupon payments c_i at T_i, $i = 1, 2, \ldots, n$ and the principal and last coupon payment will be paid back at maturity T_n. An option on this coupon bond with maturity T and strike K can be priced by applying the simulation method in the previous section together with the analytical bond price formula. All of the future cash flow of the coupon bond discounted back to time T is

$$\sum_{i=1}^{n} c_i P(T, T_i) + P(T, T_n).$$

By the risk neutral pricing principle and applying formula 11.20, the price of the option is given by

$$E\left[e^{-\int_0^T r_\tau \, d\tau} \max\left(\sum_{i=1}^{n} \hat{c}_i P(T, T_i) - K, 0\right)\right]$$

$$= E\left[e^{-\int_0^T r_\tau \, d\tau} \max\left(\sum_{i=1}^{n} \hat{c}_i e^{-\int_T^{T_i} \alpha(\tau) \, d\tau} e^{C_0(T,T_i) - D(T,T_i)x_T} - K, 0\right)\right], \quad (11.22)$$

where $\hat{c}_i = c_i$ for $i = 1, 2, \dots, n - 1$ and $\hat{c}_n = c_n + 1$. The fair price of the bond option can be found by taking the following steps:

Step 1: Simulate a path of r_t for $0 \leq t \leq T$ according to the algorithm in Example 11.2.

Step 2: Calculate the discount factor $e^{-\int_0^T r_\tau \, d\tau}$ by Equation 11.9.

Step 3: Evaluate the payoff function in Equation 11.22.

Step 4: Repeat Step 1 M times.

Although we only deal with European style bond options in this section, there are also American and Bermudan bond options in the market. American bond options can be exercised at any time within the maturity period, whereas Bermudan options only allow the holder to exercise at some discrete and prespecified dates. In terms of pricing, there are other methods, such as trinomial tree, that can price European style options. American and Bermudan options are more difficult to price due to the path dependency. Simulation thus offers a simple way to price path-dependent derivatives.

Example 11.3 *Suppose the yield to maturity is parametrized as in Example 11.1 and a = 10%, σ = 1%, r_0 = 0.002. Now consider an option on a 3-year coupon bond with a notional value of $100 and option maturity of 1 year. The coupon rate is 6% and will be paid semiannually. The strike of the option is taken to be $100.*

The fair price can be found by

$$E\left[e^{-\int_0^1 r_\tau \, d\tau} \max\left(100 \sum_{i=1}^{6} 0.03 e^{-\int_1^{0.5i+1} \alpha(\tau) \, d\tau} e^{C_0(1,0.5i+1) - D(1,0.5i+1)x_1}\right.\right.$$

$$\left.\left. + 100 e^{-\int_1^4 \alpha(\tau) \, d\tau} e^{C_0(1,4) - D(1,4)x_1} - 100, 0\right)\right]$$

$$= 6.87$$

Please see the online material for the VBA codes.

Apart from using the Euler scheme to discretize the SDE of x_t in Example 11.2, we can also use the exact simulation by solving the SDE explicitly:

$$x_t = x_s e^{a(s-t)} + \sigma e^{-at} \int_s^t e^{a\tau} \, dW_\tau. \qquad (11.23)$$

This implies that we can replace Step 3 in Example 11.2 by

Step 3': Set $x_{i+1} = x_i e^{-a\Delta t} + \sqrt{\dfrac{\sigma^2(1-e^{-2a\Delta t})}{2a}} \epsilon$.

Readers can compare the performance of these two methods in the exercise.

A range accrual is a type of structured product in which the payoff is conditional on a certain index falling within a predetermined range. The number of coupons the holder can obtain is proportional to the ratio of the observed index in the range and the number of observation dates. Buyers of range accrual products usually anticipate a steady movement of the index for it to be profitable.

Consider a simple range accrual note with principal P, which depends on the 3-month yield to maturity $R(t, t + \Delta t)$, where $\Delta t = 0.25$. For simplicity, the note is assumed to have only one coupon payment g at maturity T. Cases for quarterly or semiannually coupon rates can be extended easily. Let N be the number of observation dates, and $[h_1, h_2]$ be the preset range. After $\alpha(\tau)$ is calibrated to market data, we can price the note by simulation. The fair price of the range accrual note is given by

$$\mathrm{E}\left[e^{-\int_0^T r_\tau \, d\tau} P\left(1 + \frac{g}{N} \sum_{i=1}^N 1_{\{R(t_i, t_i + \Delta t) \in [h_1, h_2]\}}\right)\right]. \qquad (11.24)$$

As $R(t_i, t_i + \Delta t)$ is the yield to maturity that will be available only at the future time t_i, we need to evaluate it from the sample paths of r_{t_i}. Express $R(t_i, t_i + \Delta t)$ in terms of r_{t_i}, and $\alpha(\tau)$ as

$$\begin{aligned}
R(t_i, t_i + \Delta t) &= -\frac{\log P(t_i, t_i + \Delta t)}{\Delta t} \\
&= \frac{1}{\Delta t}\left(\int_{t_i}^{t_i + \Delta t} \alpha(\tau) \, d\tau - C_0(t_i, t_i + \Delta t) + D(t_i, t_i + \Delta t)x_{t_i}\right).
\end{aligned}$$

Then, for each future observation date t_i, we will be able to determine whether $R(t_i, t_i + \Delta)$ falls in the range. The simulation procedure can be summarized as follows:

Step 1: Generate a sample path of r_t for $t = 0$ to $t = T$.

Step 2: Calculate the discount factor $e^{-\int_0^T r_\tau \, d\tau}$ by Equation 11.9.

Step 3: Determine the number of $R(t_i, t_i + \Delta t)$ that fall in the range.

Step 4: Evaluate the payoff.

Step 5: Repeat Step 1 for M times.

Example 11.4 *For a 1-year range accrual note with a principal of $100 and a coupon payment of 8% that depends on the 3-month yield to maturity $R(t, t + 0.25)$, assume there are 52 observation dates and the coupon is accumulated within the range of [0.002,0.008] for $R(t, t + 0.25)$, and $a = 10\%, \sigma = 1\%, r_0 = 0.002$.*

The fair price of this accrual note is given by

$$E\left[e^{-\int_0^1 r_\tau \, d\tau} 100\left(1 + \frac{0.08}{52} \sum_{i=1}^{52} 1_{\{R(i/52, i/52+0.25)\in[0.002,0.008]\}}\right)\right] = 102.7.$$

11.6 EXERCISES

1. Let $y(s)$ be a deterministic function and W_s be a Brownian motion; consider

$$I(t) = \int_0^t \delta_s \, dW_s.$$

   (a) Using Itô's lemma on e^{uI_t} and Itô's identities on Exercise 1(d) in Chapter 4, show that

$$E[e^{uI(t)}] = 1 + \frac{1}{2}u^2 \int_0^t \delta_s^2 E\left[e^{uI(s)}\right] ds.$$

   (b) Let $y = E\left[e^{uI(t)}\right]$, the moment-generating function of $I(t)$. By differentiating the aforementioned equation, derive and solve the following ordinary differential equation (ODE):

$$\frac{dy}{dt} = \frac{1}{2}u^2 \delta_t y.$$

   Show that for each t, $I(t)$ is a normal random variable with a mean of 0 and variance of $\int_0^t \delta_s^2 \, ds$ by the uniqueness of the moment-generating function.

2. Under the Hull–White model, the zero coupon bond price has the form

$$P(t, T) = e^{\alpha(t,T)+\beta(t,T)r_t}.$$

   (a) Using the Feynman–Kac formula in Exercise 5 of Chapter 5, show that $\alpha(t, T)$ and $\beta(t, T)$ satisfy the following system of ODE:

$$\frac{\partial\alpha(t, T)}{\partial t} = -\beta(t, T)\theta(t) - \frac{1}{2}\beta^2(t, T)\sigma^2,$$

$$\alpha(T, T) = 0,$$

$$\frac{\partial\beta(t, T)}{\partial t} = -a\beta(t, T) + 1,$$

$$\beta(T, T) = 0.$$

(b) Solve the system of ODE in part (a) and compare the result with the formula 11.18.

3. Suppose the yield to maturity $R(t, T)$ is parametrized as

$$
R(t, T) = \begin{cases} a_0 + b_0 T + c_0 T^2 + d_0 T^3 + e_0 T^4, & \text{for } T \in [0, T_0], \\ a_1 + b_1 T + c_1 T^2 + d_1 T^3 + e_1 T^4, & \text{for } T \in [T_0, T_1], \\ a_2 + b_2 T + c_2 T^2 + d_2 T^3 + e_2 T^4, & \text{for } T \in [T_1, T_2]. \end{cases}
$$

Find $\alpha(T)$ for $T \in [0, T_2]$ in the Hull–White model.

4. Consider the CIR model:

$$
dr = 0.1(0.05 - r)\, dt + 0.3\sqrt{r}\, dW \quad \text{and} \quad r_0 = 0.052.
$$

(a) Construct and implement a standard Monte Carlo simulation to compute the discount factor.

(b) Use the Vasicek discount factor, which corresponds to formula 11.18 for $\theta(t) \equiv c$, where c is a constant, as a control variate to improve the simulation in (a). (Hint: you may use $\sigma_{\text{Vasicek}} = \sigma_{\text{CIR}}\sqrt{r_0}$.)

(c) Compare the difference between two prices on the basis of 1,000 simulated prices.

5. Consider the Ho–Lee interest rate movement:

$$
dr = \theta(t)\, dt + \sigma\, dW, \qquad (*)
$$

where $\theta(t) = a + e^{-bt}$, $\sigma =$ constant and W is the standard Brownian motion.

(a) Provide an algorithm to compute $B(0, t)$ by discretizing $(*)$.

(b) To price a 5-year bond paying semiannual coupons, you adopted $\Delta t = 1/250$ to calculate the integration, and $M = 1,000$ to estimate each discount factor. What is the minimum size for the random sample used to compute the bond price with simulations in (a)?

(c) Express r_t in terms of a, b, σ, h and r_0 based on the algorithm in (b), where $h = \Delta t$. Hence, show that

$$
r_t = r_0 + at + \frac{1 - e^{-bt}}{b} + \epsilon\sigma\sqrt{t}, \quad \text{when} \quad h \to 0,
$$

where $\epsilon \sim N(0, 1)$.

(d) Modify the approach in Section 11.4 to derive a closed form solution for the discount factor under the Ho–Lee model.

6. The dynamic of x_t is given by

$$
dx_t = -ax_t\, dt + \sigma\, dW_t. \tag{11.25}
$$

(a) Using Itô's lemma on $e^{at}x_t$, derive Equation 11.23, then justify the equation in Step 3'.

(b) Apply the revised algorithm (using Step 3' to simulate x_t) in Example 11.3, and compare the result with the original algorithm.

The solutions and/or additional exercises are available online at http://www.sta. cuhk.edu.hk/Book/SRMS/.

12

MARKOV CHAIN MONTE CARLO METHODS

12.1 INTRODUCTION

Bayesian inference is an important area in statistics and has also found applications in various disciplines. One of the main ingredients of Bayesian inference is the incorporation of prior information via the specification of prior distributions. As information flows freely in financial markets, incorporating prior information with Bayesian ideas constitutes a natural approach. In this final chapter, we briefly introduce the essence of Bayesian statistics with reference to risk management. In particular, we discuss the celebrated Markov Chain Monte Carlo (MCMC) method in detail and illustrate its uses via a case study.

12.2 BAYESIAN INFERENCE

The essence of the Bayesian approach is to incorporate uncertainties for the unknown parameters. Predictive inference is conducted via the joint probability distribution of the parameters $\theta = (\theta_1, \theta_2, \dots, \theta_r)$, conditional on the observable data $x = (x_1, \dots, x_n)$. The joint distribution is deduced from the distribution of observable quantities via Baye's theorem. Many excellent texts have been written about the Bayesian paradigm; see, for example, DeGroot (1970), Box and Tiao (1973), Berger (1985), O'Hagan (1994), Bernardo and Smith (2000), Lee (2004), and Robert (2001), to name just a few. Tsay (2010) provides succinct introduction to Bayesian inference for time series.

Simulation Techniques in Financial Risk Management, Second Edition. Ngai Hang Chan and Hoi Ying Wong.

The observational (or sampling) distribution $f(x|\theta)$ is the likelihood function. Under the Bayesian framework, a *prior distribution* $p(\theta)$ is specified for the parameter θ. Inferences are conducted on the basis of the posterior distribution $\pi(\theta|x)$ according to the following identity:

$$\pi(\theta|x) = \frac{f(x|\theta)p(\theta)}{f(x)},$$

where $f(x)$ is the marginal density such that

$$f(x) = \int f(x|\theta)p(\theta)\, d\theta. \tag{12.1}$$

The probability density function $\pi(\theta|x)$ is known as the *posterior* density function. Because x is observed, the marginal density in Equation 12.1 is a constant. It is more convenient to express Equation 12.1 as

$$\pi(\theta|x) \propto L(\theta)p(\theta), \tag{12.2}$$

where $L(\theta) = f(x|\theta)$ is the likelihood function. One way to estimate θ is to compute the posterior mean of θ, that is,

$$\hat{\theta} = \int \theta \pi(\theta|x)\, d\theta. \tag{12.3}$$

The prior and posterior are relative to the observables. A posterior distribution conditional on x can be used as a prior for a new observation y. This process can be iterated and eventually leads to a new posterior via Baye's theorem. We illustrate this idea with a concrete example.

Example 12.1 *Suppose that we observe $x_1, \ldots, x_n$ independent random variables each $N(\mu, \sigma^2)$ with μ unknown and σ^2 known. Estimate μ in a Bayesian setting.*

The likelihood function is

$$L(\mu) = \frac{1}{(2\pi\sigma)^{n/2}} \exp\left[-\frac{1}{2\sigma^2}\sum_{i=1}^{n}(x_i - \mu)^2\right] \propto \exp\left[-\frac{n}{2\sigma^2}(\bar{x} - \mu)^2\right],$$

where $\bar{x}$ is the sample mean of the observation. It appears natural to assume that μ follows a normal distribution by specifying the prior $p(\mu) \sim N(m, \tau^2)$, where m and τ^2 are known as hyperparameters. Substituting this prior into Equation 12.2, we have

$$\pi(\mu|x) \propto \exp\left[-\frac{(\bar{x} - \mu)^2}{2\sigma^2/n}\right] \exp\left[-\frac{(\mu - m)^2}{2\tau^2}\right]$$

$$\propto \exp\left[-\frac{(\mu - m_1)^2}{2\tau_1^2}\right],$$

TABLE 12.1   Conjugate Priors

Likelihood $L(\theta)$	Conjugate Prior $p(\theta)$
Poisson $\theta = \lambda$	$G(\alpha, \beta)$
Binomial $\theta = p$	$Be(\alpha, \beta)$
Normal $\theta = \mu, \sigma^2$ known	$N(m, \tau^2)$
Normal $\theta = \sigma^2, \mu$ known	$IG(\alpha, \beta)$

where

$$m_1 = \frac{\tau^2 \bar{x} + m\sigma^2/n}{\tau^2 + \sigma^2/n} \quad \text{and} \quad \tau_1^2 = \frac{\tau^2 \sigma^2}{n\tau^2 + \sigma^2},$$

equivalently,

$$\mu \sim N(m_1, \tau_1^2).$$

The posterior mean $\hat{\mu} = E(\mu) = m_1$ is an estimate of μ given x. Note that m_1 tends to the sample mean $\bar{x}$ and τ_1^2 tends to zero as the number of observations increases. In most cases, the prior distribution plays a lesser role when the sample size is large. Another interesting observation is that the prior contains less information as τ^2 increases. When $\tau^2 \to \infty$, $p(\mu) \propto$ constant, and $\pi(\mu|x) = N(\bar{x}, \sigma^2/n)$. Such a prior is known as a noninformative prior, as it provides no information about the distribution of μ.

There are many ways to specify a prior distribution in the Bayesian setting. Some prefer noninformative priors, and others prefer priors that are analytically tractable. Conjugate priors are adopted to address the latter concern.

Given a likelihood function, the conjugate prior distribution is a prior distribution such that the posterior distribution belongs to the same class of distributions as the prior. Conjugate priors and posterior distributions are differed through hyperparameters. Example 12.1 serves as a good example. Conjugate priors facilitate statistical inferences because the posterior distributions belong to the same family as the prior distributions, which are usually in familiar forms. Moreover, updating posterior distributions with new information becomes straightforward, as only the hyperparameters have to be updated.

In the one-dimensional case, deriving conjugate priors is relatively simple when the likelihood belongs to the exponential family. Conjugacy within the exponential family is discussed in Lee (2004). Table 12.1 summarizes some of the commonly used conjugate families. Herein, Be denotes the Beta distribution, G the Gamma distribution, IG the inverse Gamma distribution, and N the Normal distribution.

12.3   SIMULATING POSTERIORS

Bayesian inference makes use of simulation techniques to estimate the parameters naturally. As shown in Equation 12.3, calculating a posterior mean is tantamount to numerically evaluating an integral. It is not surprising, therefore, that Monte Carlo simulation plays an important role. The integration in Equation 12.3 is usually an improper integral (integration over an unbounded region), which renders standard numerical techniques useless. Although numerical quadrature can be used to bypass such a difficulty in the one-dimensional case, applying quadrature in higher dimensions is far from simple. Financial modeling usually involves higher dimensions.

Monte Carlo simulation with importance sampling simplifies the computation of Equation 12.3. As it may be difficult to generate random variables from the posterior distribution $\pi(\theta|x)$ directly, we may take advantage of the fact that importance sampling enables us to compute integrations with a conveniently chosen density. Consider

$$\hat{\theta} = \int \theta \pi(\theta|x) \, d\theta = \int \frac{\theta \pi(\theta|x)}{q(\theta)} q(\theta) \, d\theta, \tag{12.4}$$

where $q(\theta)$ is a prior specified density function that can be generated easily. Drawing n random samples θ_i from $q(\theta)$, we approximate the posterior mean by

$$\hat{\theta} = \frac{1}{n} \sum_{i=1}^{n} \frac{\theta_i \pi(\theta_i|x)}{q(\theta_i)}.$$

Note that the importance sampling is not used as a variance reduction device in this case; rather, it is applied to facilitate the computation of the posterior mean. The variance of the computation can be large in some cases.

12.4 MARKOV CHAIN MONTE CARLO

One desirable feature of combining Markov chain simulation with Bayesian ideas is that the resulting method can handle high-dimensional problems efficiently. Another desirable feature is to draw random samples from the posterior distribution directly. The MCMC methods are developed with these two features in mind.

12.4.1 Gibbs Sampling

Gibbs sampling is probably one of the most commonly used MCMC methods. It is simple, intuitive, easily implemented, and designed to handle multidimensional problems. The basic limit theorem of Markov chain serves as the theoretical building block to guarantee that draws from a Gibbs sampling agree with the posterior asymptotically.

Although conjugate priors are useful in Bayesian inference, it is difficult to construct a joint conjugate prior for several parameters. For a normal distribution with both mean and variance unknown, deriving the corresponding conjugate prior can be challenging. However, conditional conjugate priors can be obtained relatively easily; see, for example, Gilks, Richardson, and Spiegelhalter (1995). Conditioning on other parameters, a conditional conjugate prior is one dimensional and has the same distributional structure as the conditional posterior.

Gibbs sampling takes advantage of this fact and offers a way to reduce a multi-dimensional problem to an iteration of low-dimensional problems. Specifically, let $x = (x_1, \ldots, x_n)$ be the data and let the distribution of each x_i be governed by r parameters, $\theta = (\theta_1, \theta_2, \ldots, \theta_r)$. For each $j = 1, \ldots, r$, specify the one-dimensional conditional conjugate prior $p(\theta_j)$ and construct the conditional posterior by means of Baye's theorem. Then iterate the Gibbs procedure as follows.

Set an initial parameter vector $(\theta_2^{(0)}, \ldots, \theta_r^{(0)})$. Update the parameters by the following procedure:

- Sample $\theta_1^{(1)} \sim p(\theta_1 | \theta_2^{(0)}, \ldots, \theta_r^{(0)}, x)$;
- Sample $\theta_2^{(1)} \sim p(\theta_2 | \theta_1^{(1)}, \theta_3^{(0)}, \ldots, \theta_r^{(0)}, x)$;
  $$\vdots \qquad \vdots$$
- Sample $\theta_r^{(1)} \sim p(\theta_r | \theta_1^{(1)}, \theta_2^{(1)}, \ldots, \theta_{r-1}^{(1)}, x)$.

This completes one Gibbs iteration, and the parameters are updated to $(\theta_1^{(1)}, \ldots, \theta_r^{(1)})$. Using these new parameters as starting values, repeat the iteration and obtain a new set of parameters $(\theta_1^{(2)}, \ldots, \theta_r^{(2)})$. Repeating these iterations M times, we get a sequence of parameter vectors $\theta^{(1)}, \ldots, \theta^{(M)}$, where $\theta^{(i)} = (\theta_1^{(i)}, \ldots, \theta_r^{(i)})$, for $i = 1, \ldots, M$. By virtue of the basic limit theorem of Markov chain, it can be shown that the Markov chain $\{\theta^{(M)}\}$ has a limiting distribution converging to the joint posterior $p(\theta_1, \theta_2, \ldots, \theta_r | x)$ when M is sufficiently large; see Tierney (1994). The number M is called the burn-in period. After simulating $\{\theta^{(M+1)}, \theta^{(M+2)}, \ldots, \theta^{(M+n)}\}$ from the Gibbs sampling, Bayesian inference can be conducted easily. For example, to compute the posterior mean, we evaluate

$$\hat{\theta}_i = \frac{1}{n} \sum_{i=1}^{n} \theta_i^{(M+i)}.$$

To acquire a clearer understanding of Gibbs sampling, consider the following example:

Example 12.2 *One of the main uses of Gibbs sampling is to generate multivariate distributions that are usually hard to simulate by standard methods. We present a simple example to generate two correlated bivariate normal random variables θ_1 and θ_2, where*

$$\begin{bmatrix} \theta_1 \\ \theta_2 \end{bmatrix} \sim N\left(\begin{bmatrix} 0 \\ 0 \end{bmatrix}, \begin{bmatrix} 1 & \rho \\ \rho & 1 \end{bmatrix} \right).$$

To use the Gibbs sampling method, we construct a Markov chain $\{\theta^{(M)}\}$ that has a limiting distribution converging to the bivariate normal distribution $p(\theta_1, \theta_2)$. The next step is to find the marginal distribution of θ_1 given the value of θ_2. By the conditional distribution formula, we have

$$p(\theta_1|\theta_2) = \frac{p(\theta_1,\theta_2)}{p(\theta_2)}$$

$$= \frac{\frac{1}{2\pi\sqrt{1-\rho^2}}\exp\left(-\frac{\theta_1^2-2\rho\theta_1\theta_2+\theta_2^2}{2(1-\rho^2)}\right)}{\frac{1}{\sqrt{2\pi}}\exp\left(-\frac{\theta_2^2}{2}\right)}$$

$$= \frac{1}{\sqrt{2\pi}\sqrt{1-\rho^2}}\exp\left(-\frac{(\theta_1-\rho\theta_2)^2}{2(1-\rho^2)}\right).$$

From the above-mentioned functional form of the distribution function, we can conclude that, given θ_2,

$$\theta_1|_{\theta_2} \sim N(\rho\theta_2, 1-\rho^2).$$

Similarly, for θ_1, we have

$$\theta_2|_{\theta_1} \sim N(\rho\theta_1, 1-\rho^2).$$

By taking the initial guess of $\theta_2^{(0)}$ to be the mean 0, the normal random variables are generated by the following steps:

Step 1: Set $i = 1$ and $\theta_2^{(0)} = 0$.
Step 2: Generate $Z_1 \sim N(0, 1)$ and set $\theta_1^{(i)} = \rho\,\theta_2^{(i-1)} + \sqrt{1-\rho^2}Z_1$.
Step 3: Generate $Z_2 \sim N(0, 1)$ and set $\theta_2^{(i)} = \rho\,\theta_1^{(i)} + \sqrt{1-\rho^2}Z_2$.
Step 4: Set $i = i+1$.
Step 5: Go to Step 2 until i equals a prespecified integer M

   Note that $\theta_2^{(i)}$ in Step 3 is updated with the new $\theta_1^{(i)}$ generated in Step 2.
   We demonstrated how to generate these random variables using the Cholesky decomposition in Chapter 6. In this example, using Cholesky is more convenient than using Gibbs sampling. Furthermore, to generate a sequence of independent bivariate normals, we would have to perform the whole procedure from the beginning again. This shows that although Gibbs sampling is powerful for dealing with high-dimensional problems, it may not be the most efficient method.

Example 12.3   *Let $x_1, \ldots, x_n$ be independent $N(\mu, \sigma^2)$ random variables with both μ and σ^2 unknown. Estimate μ and σ^2 via Gibbs sampling.*

   Recall that the conjugate prior of μ is normal for a given σ^2 and that the conjugate prior of σ^2 is inverse gamma for a given μ. Let $\mu_0 \sim N(m_0, \tau_0^2)$ and $\sigma_0^2 \sim IG(\alpha_0, \beta_0)$ be random variables drawn from the initial priors. Define μ_i and σ_i^2 to be random variables generated in the ith iteration of the Gibbs sampling procedure. The conditional posterior for μ_i can be obtained by mimicking Example 12.1. We have

$$\mu_i|_{\sigma_{i-1}^2} \sim N(m_i, \tau_i^2),$$

where

$$m_i = \frac{\tau_{i-1}^2 \bar{x} + m_{i-1}\sigma_{i-1}^2/n}{\tau_{i-1}^2 + \sigma_{i-1}^2/n} \quad \text{and} \quad \tau_i^2 = \frac{\tau_{i-1}^2 \sigma_{i-1}^2}{n\tau_{i-1}^2 + \sigma_{i-1}^2}. \tag{12.5}$$

In Question 1 at the end of this chapter, the conditional posterior for σ_i^2 is found to be $\sigma_i^2|_{\mu_i} \sim IG(\alpha_i, \beta_i)$, where

$$\alpha_i = n/2 + \alpha_{i-1} \quad \text{and} \quad \beta_i = \beta_{i-1} + \frac{1}{2}\sum_{j=1}^{n}(x_j - \mu_i)^2. \tag{12.6}$$

Hence, Gibbs sampling is implemented as follows:

Step 1: Set $i = 1$ and initial values of m_0, τ_0^2, α_0, β_0, and σ_0^2.

Step 2: Sample $\mu_i|_{\sigma_{i-1}^2} \sim N(m_i, \tau_i^2)$ and update α_i and β_i by Equation 12.6.

Step 3: Sample $\sigma_i^2|_{\mu_i} \sim IG(\alpha_i, \beta_i)$ and update m_{i+1} and τ_{i+1}^2 by Equation 12.5.

Step 4: Set $i = i + 1$.

Step 5: Go to Step 2 until i equals a prespecified integer $M + k$.

We keep the last k pairs of random variables for indices $M + 1$ to $M + k$. The estimation is achieved by taking the sample means:

$$\hat{\mu} = \frac{1}{k}\sum_{j=1}^{k}\mu_{M+j},$$

$$\widehat{\sigma^2} = \frac{1}{k}\sum_{j=1}^{k}\sigma_{M+j}^2.$$

12.4.2   Case Study: The Effect of Jumps on Dow Jones

To appreciate the usefulness of Gibbs sampling, we use it to estimate the parameters of a jump-diffusion model and examine the effect of jumps on major financial indices. Note that maximum likelihood estimation does not work for this model (Redner and Walker, 1984).

In the jump-diffusion model of Merton (1976), the dynamics of asset returns are assumed to be

$$d \log S = \mu \, dt + \sigma \, dW_t + Y \, dN_t, \tag{12.7}$$

where S is the equity price, W_t is the standard Brownian motion, N_t follows a Poisson process with an intensity λ, and Y is a normal random variable with a mean of k and variance of s^2. We assume that dW_t, dN_t, and Y are independent random variables at each time point t. This model requires the estimation of μ, σ, λ, k, and s based on observations $\{S_1, \ldots, S_n, S_{n+1}\}$, where S_i represents the equity price observed at time t_i. These prices produce n independent log-returns, which are denoted by $\{X_1, \ldots, X_n\}$ where $X_i = \log S_{i+1} - \log S_i$. With a fixed Δt, a discrete approximation to the dynamics Equation 12.7 is

$$\Delta \log S = \mu \, \Delta t + \sigma \, \Delta W_t + Y \, \Delta N_t. \tag{12.8}$$

When Δt is sufficiently small, ΔN_t is either 1, with a probability of $\lambda \Delta t$, or 0, with a probability of $1 - \lambda \Delta t$ (Fig. 12.1).

Example 12.4   *Simulate 100 sample paths from the asset price dynamics of Equation 12.7 with the parameters $\mu = 0.08, \sigma = 0.4, \lambda = 3.5, s = 0.3$, and $k = 0$. Each sample path replicates the daily log-returns of a stock over a 1-year horizon. On the basis of these 100 paths, estimate the values of μ, σ, λ, s, and k with Gibbs sampling. Compare the results with the input values.*

Simulating paths   Sample paths are simulated by assuming $n = 250$ trading days a year, so the discretization (Eq. 12.8) has $\Delta t = 1/250$. On each path, the log-asset price at each time point is generated as follows:

$$\log S_{i+1} - \log S_i = \begin{cases} \mu \, \Delta t + \sigma \sqrt{\Delta t} \, \epsilon, & \text{if} \quad U > \lambda \Delta t \\ \mu \, \Delta t + k + \sqrt{\sigma^2 \Delta t + s^2} \, \epsilon, & \text{if} \quad U \le \lambda \Delta t \end{cases},$$

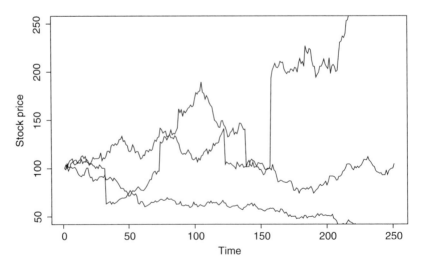

Figure 12.1   A sample path of the jump-diffusion model.

where $\epsilon \sim N(0, 1)$ and $U \sim U(0, 1)$ are independent random variables. To simplify the notations, we denote $x_i = \log S_{i+1} - \log S_i$. A graph of three sample paths is given in Figure 12.1.

Gibbs sampling   There are five parameters in the model, so we have to develop five conditional conjugate priors from their conditional likelihood functions. Let us proceed step by step.

1. Conditional prior and posterior for μ.
   Other things being fixed, the likelihood function of μ happens to be proportional to a normal density. Specifically,

$$L(\mu) \propto \prod_{i=1}^{n} \exp\left[-\frac{(x_i - \mu\Delta t - Y_i\Delta N_i)^2}{2\sigma^2\Delta t}\right]$$

$$\propto \exp\left\{-\frac{1}{2\sigma^2}\left[\mu - \sum_{i=1}^{n}(x_i - Y_i\Delta N_i)\right]^2\right\}.$$

Therefore, a normal distribution $N(m, \tau^2)$ is suitable for μ as a conditional conjugate prior. The posterior distribution can be immediately obtained as

$$N\left(\frac{\tau^2 \sum_{i=1}^{n}(x_i - Y_i\Delta N_i) + m\sigma^2/n}{\tau^2 + \sigma^2/n}, \frac{\tau^2\sigma^2}{n\tau^2 + \sigma^2}\right). \tag{12.9}$$

2. Conditional prior and posterior for σ^2.
   The conditional likelihood function of σ^2 is

$$L(\sigma^2) \propto (\sigma^2)^{-n/2} \exp\left[-\frac{1}{2\sigma^2\Delta t}\sum_{i=1}^{n}(x_i - \mu\Delta t - Y_i\Delta N_i)^2\right].$$

We select $IG(\alpha, \beta)$ as the conditional prior for σ^2. Then, the posterior distribution becomes

$$IG\left(\alpha + n/2, \beta + \frac{\sum_{i=1}^{n}(x_i - \mu\Delta t - Y_i\Delta N_i)^2}{2\Delta t}\right). \tag{12.10}$$

3. Conditional prior and posterior for λ.
   The conditional likelihood of λ is

$$L(\lambda) \propto (\lambda\Delta t)^N(1 - \lambda\Delta t)^{n-N},$$

where N is the total number of jumps in the horizon. From Table 12.1, we find that the appropriate conjugate prior is $Be(a, b)$. Simple computation shows that the posterior distribution is

$$Be\,(a + N, b + n - N). \tag{12.11}$$

4. Conditional prior and posterior for k.

   As k is the mean of the normal jump size, its prior and posterior are obtained in the same manner as μ. We state the result without proof. The prior is $N(m_Y, \tau_Y^2)$, and the posterior is given by

$$N\left(\frac{\tau_Y^2 \sum_{j=1}^{N} Y_j/N + m_Y s^2/N}{\tau_Y^2 + s^2/N}, \frac{\tau_Y^2 s^2}{N\tau_Y^2 + s^2}\right). \tag{12.12}$$

5. Conditional prior and posterior for s^2.

   As s^2 is the variance of the normal jump size, its prior and posterior are obtained in the same manner as σ^2. The prior is $IG(\alpha_Y, \beta_Y)$, and the posterior is given by

$$IG\left(\alpha_Y + N/2, \beta_Y + \frac{\sum_{i=1}^{N} (Y_i - k)^2}{2}\right). \tag{12.13}$$

The aforementioned priors and posteriors are distributions conditional on values of Y_i and ΔN_i. This complicates the Gibbs sampling procedure because only x_i is observable for all i. Therefore, at each time point t_i, Y_i, and ΔN_i should be simulated from the distributions conditional on the observed value of x_i before substituting them into the priors or posteriors. We need the following facts:

$$x_i|\Delta N_i = 0 \sim N(\mu\Delta t, \sigma^2\Delta t);$$
$$x_i|\Delta N_i = 1 \sim N(\mu\Delta t + k, \sigma^2\Delta t + s^2),$$

which together with Baye's theorem show that

$$P(\Delta N_i = 1|x_i) = \frac{P(x_i|\Delta N_i = 1)\lambda\Delta t}{P(x_i|\Delta N_i = 1)\lambda\Delta t + P(x_i|\Delta N_i = 0)(1 - \lambda\Delta t)},$$
$$P(\Delta N_i = 0|x_i) = 1 - P(\Delta N_i = 1|x_i). \tag{12.14}$$

The jump size Y_i is necessary only when $\Delta N_i = 1$. Under such a situation, we recognize that the conditional density function of Y_i is

$$f(Y_i|x_i) = f(x_i|Y_i)p(Y_i) \propto \exp\left[-\frac{(x_i - Y_i - \mu\Delta t)^2}{2\sigma^2\Delta t}\right]\exp\left[-\frac{(Y_i - k)^2}{2s^2}\right],$$

which implies

$$Y_i | x_i \sim N\left(\frac{(x_i - \mu\Delta t)/\sigma^2\Delta t + k/s^2}{1/\sigma^2\Delta t + 1/s^2}, \frac{1}{1/\sigma^2\Delta t + 1/s^2}\right). \tag{12.15}$$

With all of the ingredients ready, the Gibbs sampling starts by choosing the initial values of μ_0, σ_0^2, k_0, λ_0, and s_0^2. We also need initial values for $Y_i^{(0)}$ and $\Delta N_i^{(0)}$, both of which can be obtained by a simulation with the initial parameters. The Gibbs sampling runs as follows:

Step 1: Sample $\mu_j \sim p(\mu_j | \sigma_{j-1}^2, k_{j-1}, s_{j-1}^2, \lambda_{j-1})$, as given in Equation 12.9.

Step 2: Sample $\sigma_j \sim p(\sigma_j^2 | \mu_j, k_{j-1}, s_{j-1}^2, \lambda_{j-1})$, as given in Equation 12.10.

Step 3: Sample $\lambda_j \sim p(\lambda_j | \mu_j, \sigma_j^2, k_{j-1}, s_{j-1}^2)$, as given in Equation 12.11.

Step 4: Sample $k_j \sim p(k_j | \mu_j, \sigma_j^2, s_{j-1}^2, \lambda_j)$, as given in Equation 12.12.

Step 5: Sample $s_j^2 \sim p(s_j^2 | \mu_j, \sigma_j^2, k_j, \lambda_j)$, as given in Equation 12.13.

Step 6: Sample $\Delta N_i^{(j)} \sim p(\Delta N_i^{(j)} | \mu_j, \sigma_j^2, k_j, s_j^2)$ as given in Equation 12.14 for all $i = 1, 2, \ldots, n$.

Step 7: Sample $Y_i^{(j)} \sim p(Y_i^{(j)} | \mu_j, \sigma_j^2, k_j, s_j^2)$, as given in Equation 12.15 for the time point t_i where $\Delta N_i = 1$.

Step 8: Set $j = j + 1$ and go to Step 1. Repeat until $j = M' + M$.

Inference is drawn by taking the sample means of the values of the last M simulated parameters. The VBA code is available online in the supplementary document.

Results and comparisons   Table 12.2 shows our estimation results. We report the averaged posterior means over the 100 sample paths and the variances. As the table shows, the estimates are close to the true values, and the variances are small. Gibbs sampling does a good job of estimating the parameters for jump-diffusion models.

Example 12.4 shows the usefulness of Gibbs sampling in estimating the jump-diffusion model. In practice, this application can be crucial for a risk manager to assess how much risk is due to jumps. To examine the jump risk empirically, we estimate the effect of jumps on the Dow Jones Industrial Index. Our estimation is based on daily closing prices over the 1995-2004 period. The parameters are estimated on an annual basis.

TABLE 12.2   Performance of the Gibbs Sampling

	μ	σ^2	λ	k	s^2
True value	0.08	0.4	3.5	0	0.3
Mean	0.0769	0.3986	3.8600	0.0163	0.2868
Variance	0.0233	6.5×10^{-5}	0.8895	0.0039	0.0015

TABLE 12.3   Jump-Diffusion Estimation for Dow Jones

Year	μ	σ^2	λ	k	s^2
1995	0.2871	0.0901	1.9035	0.0627	0.2608
1996	0.2483	0.1172	2.818	−0.0337	0.235
1997	0.2384	0.1684	3.6587	−0.0256	0.2087
1998	0.1776	0.1752	5.5127	−0.0123	0.1782
1999	0.2177	0.1624	1.7968	−0.0176	0.2627
2000	−0.0162	0.1971	3.3364	−0.0235	0.2157
2001	0.015	0.1951	4.1797	−0.0383	0.2008
2002	−0.2188	0.2484	2.7072	0.0106	0.239
2003	0.1891	0.1626	2.0479	0.0661	0.2463
2004	0.0351	0.1111	1.7561	0.0004	0.2788

In Table 12.3, the number of jumps per year, λ, ranges from 1.75 to 5.5. Therefore, we can have 5–6 jumps in a particular year. The effect of jumps is significant, as almost all of the s^2 values are bigger than 0.2. The variances σ^2 associated with the Brownian motion part of the model are about 0.2 but should be divided by 250 to produce the daily variance. When a jump arrives, additional daily variance of 0.2 is added to the index return variance : $\sigma^2/250 + s^2$. The additional variance due to a jump is relatively large. Jump risk cannot be ignored! This information is useful for risk managers to construct scenarios for stress testing.

12.5   METROPOLIS–HASTINGS ALGORITHM

In this section, we explain why random draws using Gibbs sampling approximate the posterior distribution. To obtain a general result, we first introduce the Metropolis–Hastings algorithm in which the Gibbs sampling is a special case. We then show that the Metropolis–Hastings algorithm constructs a Markov chain with a limiting distribution following the posterior distribution. Further details are given in Casella and George (1992), Chib and Greenberg (1995), and Lee (2004).

Consider a Markov chain $\{\theta^{(n)}\}$ with a finite state space $\{1, 2, \ldots, m\}$ and transition probabilities p_{ij}. Given the transition probabilities, the limiting distribution of the chain can be found by solving the following equation:

$$\pi(j) = \sum_{i=1}^{m} \pi(i)p_{ij}.$$

When the state space is continuous, the sum is replaced by an integral (Bhattacharya and Waymire, 1990).

In MCMC, we work with a reverse problem. Given a posterior distribution $\pi(j)$, we want to construct a Markov chain whose transition probabilities converge to the posterior distribution. If the transition probabilities satisfy the time reversibility with respect to $\pi(j)$, then its limiting distribution is guaranteed to be equal to $\pi(j)$. To explain time reversibility, write the transition probabilities p_{ij} as

$$p_{ij} = p_{ij}^* + r_i \delta_{ij},$$

$$\delta_{ii} = 1 \quad \text{and} \quad \delta_{ij} = 0 \quad \text{for } i \neq j,$$

where $p_{ii}^* = 0$, $p_{ij}^* = p_{ij}$ for $i \neq j$, and $r_i = p_{ii}$.
  If the equation

$$\pi(i)p_{ij}^* = \pi(j)p_{ji}^* \qquad (12.16)$$

is satisfied for all i, then the probabilities p_{ij} are time reversible. This condition asserts that the probability of starting at i and ending at j when the initial probability is given by $\pi(i)$ is the same as that of starting at j and ending at i. By simple computation, we check that

$$\begin{aligned}
\sum_i \pi(i)p_{ij} &= \sum_i \pi(i)p_{ij}^* + \sum_i \pi(i)r_i\delta_{ij} \\
&= \sum_i \pi(j)p_{ji}^* + \pi(j)r_j \\
&= \pi(j)(1 - r_j) + \pi(j)r_j \\
&= \pi(j).
\end{aligned}$$

Therefore, $\pi(j)$ is the limiting distribution of the chain.
  In other words, a Markov chain whose limiting distribution is the posterior distribution can be constructed by finding a time-reversible Markov chain. We start this process by specifying the transition probabilities q_{ij}. If the probabilities q_{ij} have already satisfied the time reversibility, then the corresponding Markov chain is the one we want. Otherwise, suppose that

$$\pi(i)q_{ij} > \pi(j)q_{ji}.$$

Then, it has a higher probability of moving from i to j than from j to i. Therefore, we introduce a probability α_{ij} to reduce the moves from i to j. We would like to have

$$\pi(i)q_{ij}\alpha_{ij} = \pi(j)q_{ji}, \qquad (12.17)$$

so that

$$\alpha_{ij} = \frac{\pi(j)q_{ji}}{\pi(i)q_{ij}}.$$

As we do not want to reduce the likelihood of moving from j to i, we set $\alpha_{ji} = 1$. Therefore, the general formula is

$$\alpha_{ij} = \min\left[\frac{\pi(j)q_{ji}}{\pi(i)q_{ij}}, 1\right]. \qquad (12.18)$$

From Equations 12.17 and 12.18, we see that the transition probabilities

$$p_{ij} = q_{ij}\alpha_{ij}, \quad \text{for } i \neq j,$$
$$p_{ii} = 1 - \sum_j q_{ij}\alpha_{ij}, \tag{12.19}$$

are time reversible with respect to $\pi(i)$ and hence define a Markov chain whose limiting distribution is the required one. This method is called the Metropolis–Hastings algorithm.

Example 12.5 *Consider a random walk Markov chain:*

$$\boxed{A} \rightleftharpoons \boxed{B} \rightleftharpoons \boxed{C} \rightleftharpoons \boxed{D}$$

All transition probabilities are 0.5, except that the transitions "from A to B" and "from D to C" are 1. The transition matrix of the chain is given by

	A	B	C	D
A	0	1	0	0
B	0.5	0	0.5	0
C	0	0.5	0	0.5
D	0	0	1	0

On the basis of the Metropolis–Hastings algorithm, construct a Markov chain whose limiting distribution is $\left(\frac{1}{4}, \frac{1}{4}, \frac{1}{4}, \frac{1}{4}\right)$.

A simple calculation shows that the limiting distribution of the original Markov chain is $\left(\frac{1}{6}, \frac{1}{3}, \frac{1}{3}, \frac{1}{6}\right)$. To construct the desired Markov chain, we need to compute probabilities α_{ij}. For instance,

$$\alpha_{AB} = \min\left(1, \frac{\pi(B)\,P(A|B)}{\pi(A)\,P(B|A)}\right) = \min\left(1, \frac{\left(\frac{1}{4}\right)\left(\frac{1}{2}\right)}{\left(\frac{1}{4}\right)(1)}\right) = \frac{1}{2}.$$

This means that the transition probability "from A to B" is reduced from 1 to $1 \times \frac{1}{2} = \frac{1}{2}$. For node "A," the remaining transition probabilities correspond to the event that no transition occurs. Transition probabilities for the other nodes are obtained in the same manner. The final transition matrix becomes

	A	B	C	D
A	0.5	0.5	0	0
B	0.5	0	0.5	0
C	0	0.5	0	0.5
D	0	0	0.5	0.5

It is easy to verify that the limiting distribution of this Markov chain is $\left(\frac{1}{4}, \frac{1}{4}, \frac{1}{4}, \frac{1}{4}\right)$.

To apply the Metropolis–Hastings algorithm for simulating a random variable θ with the distribution $\pi(\theta)$, we begin with any Markov chain X_k whose transition density $q(X_k|X_{k-1})$ is easy to simulate and with a range similar to that of θ. For this Markov chain to have the desired limiting distribution $\pi(\theta)$, we need to adjust the transition density $q(X_k|X_{k-1})$ at each step k of the algorithm according to the updating criteria in Equation 12.19 so that it is time reversible. That is, if the transition probability from state X_{k-1} to state X_k is too high, we reduce its probability by α amount, then the new transition probability $p(X_k|X_{k-1})$ will form a time-reversible Markov chain with a stationary distribution of $\pi(\theta)$. The algorithm can be summarized as follows:

Step 1: Choose a transition probability q to construct the Markov chain X_k.

Step 2: Pick an initial value for θ_0 and X_0 and set $k = 1$.

Step 3: Simulate X_k according to the probability law of $q(X_k|X_{k-1})$.

Step 4: If $\alpha = \dfrac{q(X_{k-1}|X_k)\pi(X_k)}{q(X_k|X_{k-1})\pi(X_{k-1})} \geq 1$, set $\theta_k = X_k$ and go to Step 6.

Step 5: Otherwise, generate $W \sim U[0, 1]$. If $W \leq \alpha$, set $\theta_k = X_k$, otherwise set $\theta_k = \theta_{k-1}$ and $X_k = X_{k-1}$.

Step 6: Set $k = k + 1$ and repeat Step 2 until k is equal to a prespecified integer M.

Example 12.6   *In the previous chapter, we showed how to generate a normal random variable, using the acceptance-rejection method, for example. In this section, we demonstrate how a normal random variable can be generated by the Metropolis–Hastings algorithm. Let $\theta \sim N(0, 1)$. We need to construct a Markov chain that has a limiting distribution equal to a normal distribution.*

Let X_k be a stochastic process such that for each $k = 0, 1, 2, \ldots$, X_t is a double exponential random variable; that is $X_k \sim \text{DoubleExp}(1)$ with pdf, as follows:

$$p(x_k) = \frac{1}{2}e^{-|x_k|}.$$

Given the memoryless property of the double exponential,

$$P(X_{k+1}|X_k) = P(X_{k+1}),$$

it can be considered as a subclass of a Markov chain because the current state is independent of all previous states. It takes a value from negative infinity to positive infinity, making it a good candidate to approximate the normal random variable. The X_k is constructed as the initial distribution, and the transition probability will be adjusted according to the Metropolis–Hastings algorithm to transform it to a time reversible Markov chain as follows:

Step 1: Set $k = 1$, $X_0 = 0$ and $\theta_0 = 0$.

Step 2: Generate $U, V \sim U[0, 1]$.

Step 3: If $V \geq \frac{1}{2}$, set $X_k = -\ln U$, otherwise set $X = \ln U$.

Step 4: If $\alpha = \dfrac{e^{-\frac{x_k^2}{2}} \cdot e^{-|X_{k-1}|}}{e^{-\frac{x_{k-1}^2}{2}} \cdot e^{-|X_k|}} \geq 1$, set $\theta_k = X_k$ and go to Step 6.

Step 5: Otherwise, generate $W \sim U[0, 1]$. If $W \leq \alpha$, set $\theta_k = X_k$, else set $\theta_k = \theta_{k-1}$ and $X_k = X_{k-1}$.

Step 6: Set $k = k + 1$ and repeat Step 2 until k equals a prespecified integer M.

The VBA code is available online on the book's website. The following theorem justifies that the Gibbs sampling algorithm is a special case of the Metropolis–Hastings algorithm.

Theorem 12.1   *Gibbs sampling is a special case of the Metropolis–Hastings algorithm in which every jump is accepted with $\alpha \equiv 1$.*

Proof. Suppose that there are r parameters, that is, $\theta = (\theta_1, \ldots, \theta_r)$, in the model. We want to generate $\theta \sim \pi(\theta)$ for a given $\pi(\cdot)$. Let $\theta^{(0)}$ be the initial state of θ. We generate a sequence of vectors by Gibbs sampling:

$$\theta^{(0)} \to \theta^{(1)} \to \theta^{(2)} \to \theta^{(3)} \to \cdots \to \theta^{(n)} \to \theta^{(n+1)} \to \cdots,$$

where $\theta^{(n)}$ and $\theta^{(n+1)}$ only differ in one component. This sequence of vectors evolves according to the conditional density given by the Gibbs sampling algorithm. For example, the transition density from $\theta^{(k)}$ to $\theta^{(k+1)}$, where $k \leq r$, is governed by the conditional density $p(\theta_k | \theta_1, \theta_2, \ldots, \theta_{k-1}, \theta_{k+1}, \ldots, \theta_r)$. This is a Markov chain because the conditional density depends only on the previous state; in fact, only on $r - 1$ components, $(\theta_1, \theta_2, \ldots, \theta_{k-1}, \theta_{k+1}, \ldots, \theta_r)$, of the previous state. Now suppose that $\theta^{(n)}$ and $\theta^{(n+1)}$ differ in the first component:

$$\theta^{(n)} = \left(\theta_1^{(n)}, \theta_2^{(n)}, \theta_3^{(n)}, \ldots, \theta_r^{(n)} \right),$$

$$\theta^{(n+1)} = \left(\theta_1^{(n+1)}, \theta_2^{(n)}, \theta_3^{(n)}, \ldots, \theta_r^{(n)} \right),$$

where $\theta_1^{(n+1)}$ is drawn given $\left(\theta_2^{(n)}, \theta_3^{(n)}, \ldots, \theta_r^{(n)} \right)$. The transition density from $\theta^{(n)}$ to $\theta^{(n+1)}$ is given by the conditional probability density of $\pi(\cdot)$, given $\theta^{(n)}$ in the Gibbs sampling:

$$q\left(\theta^{(n+1)} | \theta^{(n)} \right) = q\left((\theta_1^{(n+1)}, \theta_2^{(n)}, \theta_3^{(n)}, \ldots, \theta_r^{(n)}) | (\theta_1^{(n+1)}, \theta_2^{(n)}, \theta_3^{(n)}, \ldots, \theta_r^{(n)}) \right)$$

$$= p\left(\theta_1 = \theta_1^{(n+1)} | \theta_2 = \theta_2^{(n)}, \theta_3 = \theta_3^{(n)}, \ldots, \theta_r = \theta_r^{(n)} \right).$$

The second equality arises because the transition density from $\theta^{(n)}$ to $\theta^{(n+1)}$ does not depend on the first component. The Metropolis–Hastings algorithm multiplies the transition density q by α, where

$$\alpha = \min \left\{ \frac{\pi\left(\theta^{(n+1)}\right) q\left(\theta^{(n)}|\theta^{(n+1)}\right)}{\pi\left(\theta^{(n)}\right) q\left(\theta^{(n+1)}|\theta^{(n)}\right)}, 1 \right\},$$

and modifies the original Markov chain to become a time-reversible one. We can prove that this Markov chain is time reversible by showing that $\alpha = 1$. Now, by conditioning on $(\theta_2, \theta_3, \ldots, \theta_r)$ with the marginal probability density $p_1(\cdot)$ of $\pi(\cdot)$, we expand $\pi\left(\theta^{(n+1)}\right)$ and $\pi\left(\theta^{(n)}\right)$ as follows:

$$\pi\left(\theta^{(n)}\right) = \pi\left(\theta_1^{(n)}, \theta_2^{(n)}, \theta_3^{(n)}, \ldots, \theta_r^{(n)}\right)$$

$$= p\left(\theta_1 = \theta_1^{(n)}|\theta_2 = \theta_2^{(n)}, \theta_3 = \theta_3^{(n)}, \ldots, \theta_r = \theta_r^{(n)}\right)$$

$$\times p_1\left(\theta_2 = \theta_2^{(n)}, \theta_3 = \theta_3^{(n)}, \ldots, \theta_r = \theta_r^{(n)}\right)$$

and

$$\pi\left(\theta^{(n+1)}\right) = \pi\left(\theta_1^{(n+1)}, \theta_2^{(n)}, \theta_3^{(n)}, \ldots, \theta_r^{(n)}\right)$$

$$= p\left(\theta_1 = \theta_1^{(n+1)}|\theta_2 = \theta_2^{(n)}, \theta_3 = \theta_3^{(n)}, \ldots, \theta_r = \theta_r^{(n)}\right)$$

$$\times p_1\left(\theta_2 = \theta_2^{(n)}, \theta_3 = \theta_3^{(n)}, \ldots, \theta_r = \theta_r^{(n)}\right).$$

Similarly, we have

$$q\left(\theta^{(n)}|\theta^{(n+1)}\right) = q\left(\left(\theta_1^{(n)}, \theta_2^{(n)}, \theta_3^{(n)}, \ldots, \theta_r^{(n)}\right) | \left(\theta_1^{(n+1)}, \theta_2^{(n)}, \theta_3^{(n)}, \ldots, \theta_r^{(n)}\right)\right)$$

$$= p\left(\theta_1 = \theta_1^{(n)}|\theta_2 = \theta_2^{(n)}, \theta_3 = \theta_3^{(n)}, \ldots, \theta_r = \theta_r^{(n)}\right).$$

The second equality is still due to the fact that from $\theta^{(n+1)}$ to $\theta^{(n)}$, the only component that differs is the first component, so the transition density is again given by the conditional density of $\pi(\cdot)$ given $(\theta_2, \theta_3, \ldots, \theta_r)$. Comparing the aforementioned four equations gives

$$\pi\left(\theta^{(n+1)}\right) q\left(\theta^{(n)}|\theta^{(n+1)}\right) = \pi\left(\theta^{(n)}\right) q\left(\theta^{(n+1)}|\theta^{(n)}\right),$$

so that $\alpha = 1$. This simply means that the probability of going from the nth state to the $(n+1)$th state is equivalent to that of going from the $(n+1)$th state to the nth state. Similarly, we can show that $\alpha = 1$ from any $\theta^{(n)}$ to $\theta^{(n+1)}$, with the kth component as the differing component. This shows that the Gibbs algorithm is indeed a particular case of Metropolis–Hastings with every jump accepted. □

When the conditional distribution of some parameters is not known explicitly, we cannot use Gibbs sampling to update the parameters, but we can still use the Metropolis–Hastings algorithm to estimate them. The following example demonstrates the use of Metropolis–Hastings in a discrete stochastic volatility model.

Example 12.7   *In the following example, we present a case study on a simple discrete stochastic volatility (SV) model by using MCMC technique to estimate the model parameters.*

Let $y_t = \log S_t - S_{t-1}$ be the difference of the log-return of stock price between time $t - 1$ and t, h_t be the log-volatility at time t, and $t = 1, 2, \ldots, n$, where n is the number of observation. Denote $y = (y_1, y_2, \ldots, y_n)$ and $h = (h_1, h_2, \ldots, h_n)$. We assume the model follows:

$$y_t = \sqrt{e^{h_t}}\, \epsilon_t, \tag{12.20}$$

$$h_{t+1} = \mu + \tau \eta_t, \tag{12.21}$$

where $h_1 \sim N(\mu, \tau^2)$. ϵ_t and η_t are assumed to be independent and follow normal distribution with mean 0 and variance 1 as follows

$$\begin{bmatrix} \epsilon_t \\ \eta_t \end{bmatrix} \sim N\left(\begin{bmatrix} 0 \\ 0 \end{bmatrix}, \begin{bmatrix} 1 & 0 \\ 0 & 1 \end{bmatrix} \right),$$

for all $t \in \mathbb{N}$.

To sample the parameters, one of the possible ways is to perform the Gibbs sampling algorithm as follows:

Step 1: Initialize $h^{(0)}$, τ_0^2 and μ_0 and set $i = 1$.
Step 2: For $t = 1, \ldots, n$, sample $h_t^{(i)} \sim p\left(h_t^{(i)} | \mu_{i-1}, \tau_{i-1}^2, y, h_{>t}^{(i-1)}, h_{<t}^{(i)} \right)$, where $h_{>t}^{(i-1)} = h_{t+1}^{(i-1)}, \ldots, h_n^{(i-1)}$ and $h_{<t}^{(i)} = h_1^{(i)}, \ldots, h_{t-1}^{(i)}$.
Step 3: Sample $\mu_i \sim p\left(\mu_i | \tau_{i-1}^2, y, h^{(i)} \right)$.
Step 4: Sample $\tau_i^2 \sim p\left(\tau_i^2 | \mu_i, y, h^{(i)} \right)$.
Step 5: Repeat Step 2 by setting $i = i + 1$ for M times.

By Baye's rule, we can derive the conditional posteriors as follows

$$p(\mu | \tau^2, y, h) \propto p(y|h)\, p(h|\mu, \tau^2)\, p(\mu) \quad \text{and}$$

$$p(\tau^2 | \mu, y, h,) \propto p(y|h)\, p(h|\mu, \tau^2)\, p(\tau^2),$$

where $p(\mu)$ and $p(\tau^2)$ are independent priors. In this case, we take $p(\mu) \sim N(\alpha_\mu, \beta_\mu)$ and $p(\tau^2) \sim IG(\alpha_\tau, \beta_\tau)$, where $IG(\cdot, \cdot)$ denotes the inverse gamma distribution, $\alpha_{(\cdot)}$

and $\beta_{(\cdot)}$ are hyperparameters specified by users. To obtain the conditional posterior distribution for μ, we apply Baye's rule as follows

$$p(\mu | \tau^2, y, h) \propto p(h | \mu, \tau^2) p(\mu)$$

$$\propto \prod_{t=1}^{n} p(h_t | \mu, \tau^2) N(\alpha_\mu, \beta_\mu)$$

$$\propto \exp\left\{ -\frac{1}{2\tau^2} \left[\sum_{t=1}^{n} (h_t - \mu)^2 \right] \right\} \exp\left\{ -\frac{(\mu - \alpha_\mu)^2}{2\beta_\mu^2} \right\}$$

$$\propto \exp\left\{ -\frac{(\mu - \hat{\alpha}_\mu)^2}{2\hat{\beta}_\mu^2} \right\},$$

where

$$\hat{\alpha}_\mu = \frac{\bar{h}\beta_\mu^2 + \alpha_\mu \tau^2 / n}{\beta_\mu^2 + \tau^2 / n}, \quad \hat{\beta}_\mu = \frac{\beta_\mu^2 \tau^2}{n\beta_\mu + \tau^2}, \quad \text{and} \quad \bar{h} = \frac{1}{n} \sum_{t=1}^{n} h_t.$$

Similarly, the conditional posterior distribution for τ^2 can be obtained as follows

$$p(\tau^2 | \mu, y, h,) \propto p(h | \mu, \tau^2) p(\tau^2)$$

$$\propto \left(\frac{1}{\tau^2}\right)^{n/2} \prod_{t=1}^{n} p(h_t | \mu, \tau^2) IG(\alpha_\tau, \beta_\tau)$$

$$\propto \left(\frac{1}{\tau^2}\right)^{n/2} \exp\left\{ -\frac{1}{2\tau^2} \left[\sum_{t=1}^{n} (h_t - \mu)^2 \right] \right\} \frac{(\beta_\tau)^{\alpha_\tau} e^{-\beta_\tau / \tau^2}}{\Gamma(\alpha_\tau)(\tau^2)^{\alpha_\tau + 1}}$$

$$\propto \exp\left\{ -\frac{1}{\tau^2} \left[\beta_\tau + \frac{1}{2} \sum_{t=1}^{n} (h_t - \mu)^2 \right] \right\} \left(\frac{1}{\tau^2}\right)^{(\alpha_\tau + n/2) + 1}$$

$$\propto IG(\hat{\alpha}_\tau, \hat{\beta}_\tau),$$

where

$$\hat{\alpha}_\tau = \alpha_\tau + \frac{n}{2} \quad \text{and} \quad \hat{\beta}_\tau = \beta_\tau + \frac{1}{2} \sum_{t=1}^{n} (h_t - \mu)^2.$$

To sample h_t from $p(h_t | \mu, \tau, y, h_{-t})$, we first derive its conditional posterior distribution as follows

$$p(h_t | \mu, \tau, y, h_{-t}) \propto p(y_t | \mu, \tau^2, h_t) p(h_t | \mu, \tau^2, h_{-t})$$

$$\propto \frac{1}{\sqrt{2\pi e^{h_t}}} \exp\left\{ -\frac{y_t^2}{2e^{h_t}} \right\} \exp\left\{ -\frac{(h_t - \mu)^2}{2\tau^2} \right\}.$$

This density function is not easy to sample directly. One can use the acceptance-rejection method in Chapter 6 by finding a density $q(h_t)$ and a constant c such that $p(h_t) \leq cq(h_t)$ for simulating the conditional posterior distribution. The Metropolis–Hastings algorithm provides an alternative way to sample this density easily. Let X_t be a Markov chain with transition density as the normal density $q(x_t) \propto \exp\{-(x_t - \alpha_h)^2/2\beta_h\}$ so that it does not depend on any of the previous states. Specifically, simulate a random variable x_t from $q(x)$, accept x_t as $h_t^{(i)}$ with probability

$$
\min\left(\frac{p(x_t|\mu, \tau, y, h_{-t})\, q\left(h_t^{(i-1)}\right)}{p\left(h_t^{(i-1)}|\mu, \tau, y, h_{-t}\right) q(x_t)}, 1 \right).
$$

Otherwise, set $h_t^{(i)} = h_t^{(i-1)}$ and move to sample $h_{t+1}^{(i)}$.

Different choices of $q(x)$ can lead to different efficiency of the algorithm. Interested readers may refer to Jacquier et al. (1994) for details. In practice, the log-return can be adjusted to have a mean of zero if we minus each of y_t by the mean $u = \left(\sum_{t=1}^{n} y_t\right)/n$. Then, the mean-corrected returns $\hat{y}_t = y_t - u$ can be applied directly in this simple SV model. Some software packages, such as WINBUS, can perform the sampling conveniently for users.

12.6   EXERCISES

1. Suppose that $X_1, \ldots, X_n$ are independent observations that follow $N(\mu, \sigma^2)$, where μ is a known quantity.

   (a) Show that the likelihood function $L(\sigma^2)$ satisfies

   $$
   L(\sigma^2) \propto (\sigma^2)^{-n/2} \exp\left\{ -\frac{1}{2} \sum_{i=1}^{n} (X_i - \mu)^2 \right\}.
   $$

   (b) Suppose further that $\sigma^2 \sim IG(\alpha, \beta)$. What is the conditional distribution of $\sigma^2|X_1, \ldots, X_n$?
   Hint: Denote $p(\phi)$ as the density of the inverse Gamma distribution. Then we have

   $$
   p(\phi) \propto \phi^{-\alpha+1} e^{-\beta/\phi}.
   $$

2. A density function with a single parameter, $p(x|\theta)$, is said to be of the exponential family if it takes the form

   $$
   p(x|\theta) = g(x)h(\theta) \exp\left[\sum t(x)\psi(\theta) \right].
   $$

   Show that a normal mean with a known variance, normal variance with a know mean, a Poisson distribution, and a binomial distribution are of the exponential family.

3. Show that if the likelihood function is from the exponential family and the prior distribution is from the exponential family, then the posterior distribution also belongs to the exponential family.

4. Simulate the daily jump-diffusion VaR (value-at-risk) of the Dow Jones Industrial Index on the basis of the data used in Section 7.2.3. Compare your value with the GED-VaR (generalized error distribution–value-at-risk) defined in Chapter 7.

5. Suppose that $x|p \sim Bin(n, p)$ and $p|x \sim Be(x + \alpha, n - x + \beta)$, where n is a Poisson variable of mean λ. Use Gibbs sampling to find the unconditional distribution of n where $\lambda = 16$, $\alpha = 2$ and $\beta = 4$.

6. Consider the normal distribution with an unknown mean μ and a known variance.
   (a) Assume that the prior of μ is a discrete mixture of two normal densities. Show that this prior is still conjugate.
   (b) Assume that the prior of μ is a discrete mixture of k normal densities. Is the prior still conjugate?

7. Consider the following transition matrix of a Markov chain:

	1	2	3	4
1	1/6	0	1/2	1/3
2	0	1/3	1/3	1/3
3	0	1/2	0	1/2
4	1/4	1/4	1/4	1/4

   Use the Metropolis–Hastings algorithm to construct a Markov chain whose limiting distribution is $(1/6, 1/6, 1/3, 1/3)$ based on the aforementioned matrix.

8. Consider the transition matrix of another Markov chain:

	1	2	3	4
1	1/2	0	1/2	0
2	2/3	0	1/6	1/6
3	0	1/3	0	2/3
4	1/4	1/4	1/4	1/4

   Use the Metropolis–Hastings algorithm to construct a Markov chain whose limiting distribution is $(1/10, 2/10, 3/10, 4/10)$ based on the aforementioned matrix.

9. Modify the online VBA supplementary codes from Example 12.6 to generate the GED (see Section 7.2.2 for the details of this distribution). By choosing $\xi = 1.6$, compare the shape of the generated distribution to that of the standard normal distribution (which corresponds to $\xi = 2$ in the GED distribution) and the double exponential distribution (which corresponds to $\xi = 1$ in the GED distribution).

The solutions and/or additional exercises are available online at http://www.sta.cuhk.edu.hk/Book/SRMS/.

REFERENCES

1. Alexander, C.A. (ed.) (1998a). *Risk Management and Analysis, Volume 1: Measuring and Modeling Financial Risk*. John Wiley & Sons, Ltd, Chichester.

2. Alexander, C.A. (ed.) (1998b). *Risk Management and Analysis, Volume 2: New Markets and Products*. John Wiley & Sons, Ltd, Chichester.

3. Anderson, T.W. (2003). *An Introduction to Multivariate Statistical Analysis*, 3rd ed. John Wiley & Sons, Inc., New York.

4. Barnett, V. (1991). *Sample Surveys: Principles and Methods*, 2nd ed. Oxford University Press, New York.

5. Berger, J.O. (1985) *Statistical Decision Theory and Bayesian Analysis*, 2nd ed. Springer-Verlag, New York.

6. Bernardo, J.M. and Smith, A.F.M. (2000). *Bayesian Theory*. John Wiley & Sons, Ltd, Chichester.

7. Bhattacharya, R.N. and Waymire, E.C. (1990). *Stochastic Processes with Applications*. John Wiley & Sons, Inc., New York.

8. Billingsley, P. (1999). *Convergence of Probability Measures*, 2nd ed. John Wiley & Sons, Inc., New York.

9. Black, F. and Scholes, M. (1973). The pricing of option and corporate liabilities. *Journal of Political Economy* **81**, 637–659.

10. Box, G.E.P. and Tiao, G.C. (1973). *Bayesian Inference in Statistical Analysis*. Addison-Wesley, Reading, Massachusetts.

11. Brandimarte, P. (2006). *Numerical Methods in Finance and Economics: A MATLAB-Based Introduction,* 2nd ed. John Wiley & Sons, Inc., New York.

12. Broadie, M. and Glasserman, P. (1998). Simulation for option pricing and risk management. In Alexander, C. (ed.), *Risk Management and Analysis, Volume 1: Measuring and Modeling Financial Risk*, 173–206. John Wiley & Sons, Ltd, Chichester.

13. Casella, G. and Berger, R.L. (2001). *Statistical Inference*, 2nd ed. Duxbury Press, Belmont, California.

14. Casella, G. and Goerge, E. (1992). Explaining the Gibbs sampler. *American Statistician* **46**, 167–174.

15. Chan, N.H. (2010). *Time Series: Applications to Finance with R and S-Plus,* 2nd ed. John Wiley & Sons, Inc., New York.

16. Chan, N.H. and Wong, H.Y. (2013). *Handbook of Financial Risk Management: Simulations and Case Studies*. John Wiley & Sons, Inc., Hoboken, New Jersey.

17. Chib, S. and Greenberg, E. (1995). Understanding the Metropolis-Hastings algorithm. *American Statistician* **49**, 327–335.

18. Conte, S.D. and de Boor, C. (1980). *Elementary Numerical Analysis: An Algorithmic Approach*, 3rd ed. McGraw-Hill, New York.

19. Cox, J., Ross, A., and Rubinstein, M. (1979). Option pricing: a simplified approach. *Journal of Financial Economics* **7**, 229–264.

20. Cox, J., Ingersoll, J., and Ross, A. (1985). A theory of the term structure of interest rates. *Econometrica* **53**, 385–407.

21. Crouchy, M., Galai, D., and Mark, R. (2000). *Risk Management*. McGraw-Hill, New York.

22. Dana, R.A. and Jeanblanc, M. (2002). *Financial Markets in Continuous Time*. Springer-Verlag, Berlin.

23. DeGroot, M.H. (1970). *Optimal Statistical Decisions*. McGraw-Hill, New York.

24. Dempster, M.A.H. (ed.) (2002). *Risk Management: Value at Risk and Beyond*. Cambirdge University Press, Cambridge.

25. Embrechts, P., Klüppelberg, C., and Mikosch, T. (1997). *Modelling Extremal Events for Insurance and Finance*. Springer-Verlag, Berlin.

26. Finkenstädt, B. and Rootzén, H. (eds.) (2004). *Extreme Values in Finance, Telecommunications and the Environment*. Chapman and Hall, New York.

27. Fournie, E., Lasry, J., Lebuchoux, J., Lions, P., and Touzi, N. (1999). Applications of Malliavin Calculus to Monte Carlo methods in finance. *Finance and Stochastics* **3**, 391–412.

28. Fournie, E., Lasry, J., Lebuchoux, J., and Lions, P. (2000). Applications of Malliavin calculus to Monte Carlo methods in finance II. *Finance and Stochastics* **5**, 201–236.

29. Gelman, A., Carlin, J.B., Stern, H.S., and Rubin, D.B. (eds.) (2003). *Bayesian Data Analysis*, 2nd ed. Chapman and Hall, New York.

30. Gilks, W.R., Richardson, S., and Speigelhalter, D.J. (eds.) (1995). *Markov Chain Monte Carlo in Practice*. Chapman and Hall, New York.

31. Glasserman, P. (2003). *Monte Carlo Methods in Financial Engineering*. Springer-Verlag, New York.

32. Hansen, A.T. and Jorgensen, P.L. (2000). Analytical valuation of American-style Asian options. *Management Science* **46**, 1116–1136.

33. Heath, D., Jarrow, R., and Morton, A. (1992). Bond pricing and the term structure of interest rates. *Econometrica* **60**, 77–106.

34. Hickernell, F.J., Lemieux, C., and Owen, A.B. (2005). Control variates for Quasi-Monte Carlo (with discussions). *Statistical Science* **20**, 1–31.

35. Ho, T. and Lee, S. (1986). Term structure movements and pricing interest rate contingent claims. *Journal of Finance* **41**, 1011–1029.

36. Hogg, R.V. and Tanis, E.A. (2006). *Probability and Statistical Inference*, 7th ed. Prentice Hall, Englewood Cliffs, New Jersey.

37. Hull, J. (2006). *Options, Futures and Other Derivatives*, 6th ed. Prentice Hall, Englewood Cliffs, New Jersey.

38. Hull, J. and White, A. (1988). The use of the control variate technique in option pricing. *Journal of Financial and Quantitative Analysis* **23**, 237–251.

39. Hull, J. and White, A. (1994). *Numerical procedures for implementing term structure models I: Single-factor models.* Journal of Derivatives **2**, 7–16.

40. Jacquier, E., Polson, N., and Rossi, P. (1994). Bayesian analysis of stochastic volatility models (with discussion). *Journal of Business and Economic Studies* **12**(4), 371-417.

41. Jaeckel, P. (2002). *Monte Carlo Methods in Finance*. John Wiley & Sons, Inc., New York.

42. Jarrow, R.A. (2002). *Modeling Fixed-Income Securities and Interest Rate Options*. Stanford University Press, Stanford, California.

43. Jorion, P. (2000). *Value at Risk*, 2nd ed. McGraw-Hill, New York.

44. Joshi, M. (2003). *The Concepts and Practice of Mathematical Finance*. Cambridge University Press, Cambridge.

45. Karatzas, I. and Shreve, S.E. (1997). *Brownian Motion and Stochastic Calculus*, 2nd ed. Springer-Verlag, New York.

46. L'Ecuyer, P. (1994). Uniform random number generation. *Annals of Operations Research* **53**, 77–120.

47. Lee, P. (2004). *Bayesian Statistics: An Introduction*, 3rd ed. Edward Arnold, London.

48. Longstaff, F. and Schwartz, E.S. (2001). Valuing American options by simulation: a simple least-squares approach. *Review of Financial Studies* **14**, 113–147.

49. Margrabe, W. (1978). The value of an option to exchange one asset for another. *Journal of Finance* **33**, 177–186.

50. McNeil, A.J., Frey, R., and Embrechts, P. (2005). *Quantitative Risk Management*. Princeton University Press, Princeton, New Jersey.

51. Merton, R.C. (1973). Theory of rational option pricing. *Bell Journal of Economics and Management Science* **4**, 141–183.

52. Merton, R.C. (1976). *Option pricing when underlying stock returns are discontinuous.* Journal of Financial Economics **3**: 125–144.

53. Moreno, M. and Navas, J.F. (2003). On the robustness of least-squares Monte Carlo for pricing American derivatives. *Review of Derivatives Research* **6**, 107–128.

54. O'Hagan, A. (1994). *Kendall's Advanced Theory of Statistics, Volume 2B: Bayesian Inference*. Edward Arnold, London.

55. Pliska, S.R. (1997). *Introduction to Mathematical Finance: Discrete Time Models*. Blackwell, Maiden, Massachusetts.

56. Redner, R. and Walker, H. (1984). Mixture densities, maximum likelihood and the EM algorithm. *SIAM Review* **26**, 195–239.

57. Robert, C.P. (2001). *The Bayesian Choice: From Decision-Theoretic Foundations to Computational Implementation*, 2nd ed. Springer-Verlag, New York.

58. Ross, S. (2002). *Simulation*, 3rd ed. Academic Press, San Diego, California.

59. Stentoft, L. (2004). Assessing the least squares Monte-Carlo approach to American option valuation. *Review of Derivatives Research* **7**, 129–168.

60. Tierney, L. (1994). Markov chains for exploring posterior distributions, with discussions. *Annals of Statistics* **22**, 1701–1762.

61. Tsay, R.S. (2010). *Analysis of Financial Time Series*, 3rd ed. John Wiley & Sons, Inc., New York.

62. Vasicek, O. (1977). An equilibrium characterization of the term structure. *Journal of Financial Economics* **5**, 177–188.

63. Weisberg, S. (1985). *Applied Linear Regression*, 2nd ed. John Wiley & Sons, Inc., New York.

64. Wong, H.Y. and Cheung, Y.L. (2004). Geometric Asian options: valuation and calibration with stochastic volatility. *Quantitative Finance* **4**, 301–314.

65. Wong, H.Y. and Kwok, Y.K. (2003). Sub-replication and replenishing premium: efficient pricing of multi-state lookbacks. *Review of Derivatives Research* **6**, 83–106.

INDEX

Simulation Techniques in Financial Risk Management, Second Edition. Ngai Hang Chan and Hoi Ying Wong.
© 2015 John Wiley & Sons, Inc. Published 2015 by John Wiley & Sons, Inc.

WILEY SERIES IN STATISTICS IN PRACTICE

Advisory Editor, MARIAN SCOTT, *University of Glasgow, Scotland, UK*

Founding Editor, VIC BARNETT, *Nottingham Trent University, UK*

Human and Biological Sciences

Brown and Prescott · Applied Mixed Models in Medicine
Ellenberg, Fleming and DeMets · Data Monitoring Committees in Clinical Trials:
  A Practical Perspective
Lawson, Browne and Vidal Rodeiro · Disease Mapping With WinBUGS and
  MLwiN
Lui · Statistical Estimation of Epidemiological Risk
*Marubini and Valsecchi · Analysing Survival Data from Clinical Trials and
  Observation Studies
Parmigiani · Modeling in Medical Decision Making: A Bayesian Approach
Senn · Cross-over Trials in Clinical Research, *Second Edition*
Senn · Statistical Issues in Drug Development
Spiegelhalter, Abrams and Myles · Bayesian Approaches to Clinical Trials and
  Health-Care Evaluation
Turner · New Drug Development: Design, Methodology, and Analysis
Whitehead · Design and Analysis of Sequential Clinical Trials, *Revised Second
  Edition*
Whitehead · Meta-Analysis of Controlled Clinical Trials
Zhou, Zhou, Liu and Ding · Applied Missing Data Analysis in the Health Sciences

Earth and Environmental Sciences

Buck, Cavanagh and Litton · Bayesian Approach to Interpreting Archaeological
  Data
Cooke · Uncertainty Modeling in Dose Response: Bench Testing Environmental
  Toxicity
Gibbons, Bhaumik, and Aryal · Statistical Methods for Groundwater Monitoring,
  Second Edition
Glasbey and Horgan · Image Analysis in the Biological Sciences
Helsel · Nondetects and Data Analysis: Statistics for Censored Environmental Data
Helsel · Statistics for Censored Environmental Data Using Minitab® and R,
  Second Edition
McBride · Using Statistical Methods for Water Quality Management: Issues,
  Problems and Solutions
Ofungwu · Statistical Applications for Environmental Analysis and Risk Assessment
Webster and Oliver · Geostatistics for Environmental Scientists

Industry, Commerce and Finance

Aitken and Taroni · Statistics and the Evaluation of Evidence for Forensic Scientists,
  Second Edition
Brandimarte · Numerical Methods in Finance and Economics: A MATLAB-Based
  Introduction, *Second Edition*

*Now available in paperback.

Brandimarte and Zotteri · Introduction to Distribution Logistics

Chan and Wong · Simulation Techniques in Financial Risk Management, *Second Edition*

Jank · Statistical Methods in eCommerce Research

Jank and Shmueli · Modeling Online Auctions

Lehtonen and Pahkinen · Practical Methods for Design and Analysis of Complex Surveys, *Second Edition*

Lloyd · Data Driven Business Decisions

Ohser and Mücklich · Statistical Analysis of Microstructures in Materials Science

Rausand · Risk Assessment: Theory, Methods, and Applications